This work is supported by China-Poland Inter-Governmental Science and Technology Cooperation Project (2012-35-5), Wuxi City Vocational and Technical College.

基于GIS的决策支持与表面分析
GIS–Based Decision Support & Surface Analysis

陈　静　朱庆杰　著

吉林大学出版社

· 长春 ·

图书在版编目（CIP）数据

基于GIS的决策支持与表面分析：英文 / 陈静，朱庆杰著. —长春：吉林大学出版社，2023.5
ISBN 978-7-5768-1644-0

Ⅰ．①基… Ⅱ．①陈… ②朱… Ⅲ．①地理信息系统 Ⅳ．①P208.2

中国国家版本馆CIP数据核字（2023）第079218号

书　　名：基于GIS的决策支持与表面分析
GIS-Based Decision Support & Surface Analysis

作　　者：陈　静　朱庆杰
策划编辑：邵宇彤
责任编辑：刘　丹
责任校对：赵　莹
装帧设计：优盛文化
出版发行：吉林大学出版社
社　　址：长春市人民大街4059号
邮政编码：130021
发行电话：0431-89580028/29/21
网　　址：http://www.jlup.com.cn
电子邮箱：jldxcbs@sina.com
印　　刷：三河市华晨印务有限公司
成品尺寸：170mm×240mm　　16开
印　　张：11.75
字　　数：185千字
版　　次：2023年5月第1版
印　　次：2023年5月第1次
书　　号：ISBN 978-7-5768-1644-0
定　　价：78.00元

版权所有　　翻印必究

Abstract

In this book, we mainly focus on decision support and surface analysis of GIS analysis in IDRISI. In decision support, multi-criteria and multi-objective decision making are analyzed, such as Boolean approach, weighted linear combination, ordered weighted averaging and multiple objectives. Also, weights calculation method, analysis procedure and application examples are investigated. In surface analysis, tin and tin surface, dempster-shafer theory, spatial dependence modeler, model fitting and ordinary kriging method are analyzed; and the core of this part is spatial variability analysis. Several application examples in suitability and risk analysis are introduced, such as air pollution analysis, site evaluation in land use, casing failure and reservoir heterogeneity analysis in petroleum engineering. The contents are described according to the actual operation procedure, thus those methods can be used by readers conveniently. This book can be used as a reference book for student teaching, also for scientific and technical staff, which cares about disaster reduction, safety evaluation and GIS application.

Contents

Part I Decision Support ·········· 001

    Chapter 1   Multi-Criteria Evaluation ·········· 003

        1.1 Boolean Intersection and Weighted Linear Combination ······ 003

        1.2 Criteria Weights Calculation ·········· 008

        1.3 Application Examples ·········· 012

    Chapter 2   Ordered Weighted Averaging ·········· 033

        2.1 Method Introduction ·········· 035

        2.2 Order Weights Calculation ·········· 037

        2.3 Application Examples ·········· 049

    Chapter 3   Multiple Objectives ·········· 073

        3.1 Conflict Objectives Analysis ·········· 073

        3.2 Application Examples ·········· 079

Prat II Surface Analysis ·········· 095

    Chapter 4   Uncertainty Analysis ·········· 097

        4.1 Dempster-Shafer Theory ·········· 097

        4.2 An Application Example ·········· 102

    Chapter 5   Geostatistics ·········· 113

        5.1 TIN and TIN Surface ·········· 113

5.2 Spatial Dependence Modeler ············ 119
5.3 Model Fitting ············ 122
5.4 Ordinary Kriging ············ 124
Chapter 6　Application Examples of Geostatistics ············ 129
6.1 Application in Air Pollution ············ 129
6.2 Application in Casing Failure Analysis ············ 150
6.3 Application in Reservoir Heterogeneity ············ 165

References ············ 175

Part I Decision Support

Chapter 1 Multi-Criteria Evaluation

Multi-Criteria Decision Making in GIS is one of the important parts of decision making, in which includes Multi-Criteria Evaluations for a single objective and Multi-Objective Evaluations for several objectives. Multi-Criteria Evaluation is a procedure to meet one specific decision making target (objective) with several evaluated factors (criteria). Therefore, Multi-Criteria Evaluation (MCE) can be explained as a method to obtain evaluation result through combining several factors (criteria) to a single index. In IDRISI, three methods are provided in MCE module to combine multiple criteria; they are Boolean Intersection (BI), Weighted Linear Combination (WLC), and Ordered Weighted Average (OWA).

1.1 Boolean Intersection and Weighted Linear Combination

The simplest method of MCE decision making procedure is Boolean overlay, in which suitable or unsuitable value is represented by 1 or 0. It means that 1 represents suitable and 0 for unsuitable. In other words, only suitable or unsuitable can be selected, and no other choices. The combination (or aggregation) of criteria is fulfilled with logical Boolean operators intersection (AND) and union (OR). It is an extreme decision-making form. For Boolean Intersection (logical AND), all criteria must be met in the decision set, it will be excluded if any one criterion fails to be met; and for Boolean Union (logical OR), only a single criterion is needed to be met. In Boolean Intersection, the worst quality must meet the decision objective; therefore, it is risk-averse. On the other hand, Boolean Union is opposite, it is risk involved, and only a single criterion is needed to be met. Boolean Intersection is integrated within the MCE module.

The second method of MCE decision making procedure is known as Weighted Linear Combination (WLC) that is most commonly used; it is a weighted average method with standardized continuous factors. It's between extremes AND and OR, also, neither risk involved nor risk averse. In the procedure of WLC, each factor is standardized and multiplied by its factor weight, then sums all factors. Since all factor weights sum to one, the evaluation result has the same range of values as every standardized factor. For example, for any spatial point (pixel) (x, y) with m factors, its suitability can be expressed as,

$$f(x,y) = \sum_{i=1}^{m} u_i z_i(x,y) \qquad (1-1)$$

In which,

$f(x, y)$ = suitability at a spatial point (x, y),

u_i = criterion weight of factor i,

$z_i(x, y)$ = standardized criterion score of factor i at point (x, y).

The result is a continuous suitable image. The evaluation is an aggregation procedure of factors (influence factors) based on criteria weights and standardized factor images (criteria maps). Factors are decision variables that are measured on a continuous scale and also treated as criteria. Therefore, a factor is also called a criterion.

A criterion is an individual evidence for a decision that can be assigned to a decision set, measured and evaluated. In IDRISI, there are two kinds of criteria, factors and constraints, factors are continuous criteria, and constraints are Boolean criteria. Since criteria are measured with different scales, it is necessary to standardized all factors for Formula (1-1).

This is called standardization, linear scaling is the simplest standardization, and can be expressed as,

$$z_i(x,y) = \frac{R_i(x,y) - R_{i\min}}{R_{i\max} - R_{i\min}} \quad (1-2)$$

In which,

$R_i(x,y)$ = raw score of factor i at a spatial point (x, y),

$R_{i\max}$ = maximum raw score of factor i,

$R_{i\min}$ = minimum raw score of factor i.

This is a linear membership function; it can be treated as the simplest one of fuzzy membership functions. Many fuzzy sets membership functions are offered to standardize factors by FUZZY module in IDRISI, in which continuous factors are considered as fuzzy sets, either a 0–1 real number scale or a 0–255 byte scale is optional. See Figure 1–1.

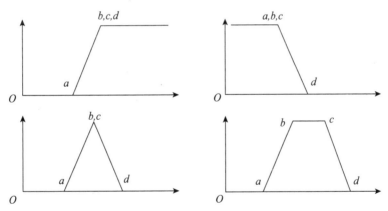

Fig.1-1 Common membership functions

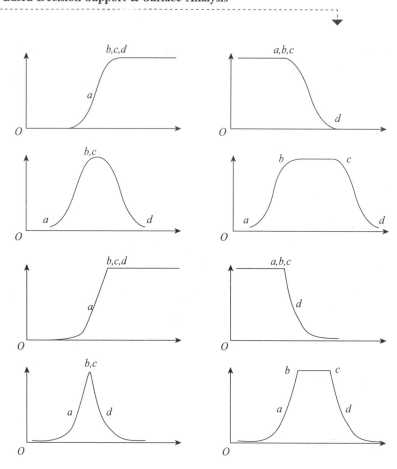

Fig.1-1　Common membership functions (continued)

It is worth mentioning about critical points for the set membership function that have a tremendous impact on the evaluation. For example, if the influence range of the fault zone to buildings is 2 km, it means that all buildings far away from the fault zone 2 km are safe. Thus, the critical point is 2 km from fault zones. With this critical point, a location 10 km from the fault zone has the same standardized value 255 for a byte scaling (or 1.0 for a real scaling) with the location 2 km. Otherwise, according to Formula (1-2), if the value of 10 km away point from the fault zone is 1.0, then the point with 2 km would be 0.2. When we develop a fuzzy membership function, the critical points or the end points should be considered carefully.

If Boolean constraints are also applied in Formula (1-1), it can be shown as,

$$f(x,y) = \sum_{i=1}^{n} u_i z_i(x,y) \times \prod c_j(x,y) \tag{1-3}$$

In which,

c_j = criterion score of constraint j at point (x,y),

$\prod$ = product.

Formula (1-3) is the modification of Formula (1-1) for the evaluation procedure through multiplying the constraints. Constraints are limiting factors that are usually expressed as logical images (Boolean Intersection or union maps) to limit decision scheme under consideration. The inherent meaning is that a continuous suitability image can be masked by Boolean constraints.

All those options can be fulfilled by WLC in MCE module in IDRISI (see Figure 1-2). In WLC option, you should input the number of factors, file names, and criteria weights, the number of constraints and file names. If constraints are specified, the output suitability image will be masked by logical images. For example, unsuitable areas for building can be used as the constraints to modify (mask) evaluation result (suitability image) for site selection. Of course, all options in MCE module can be fulfilled with Image Calculator, decision wizard or a combination of SCALAR and OVERLAY.

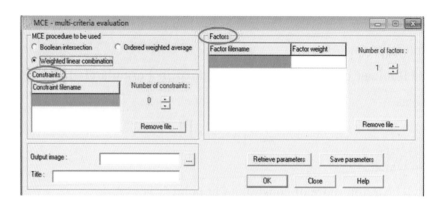

Fig.1-2　WLC option in MCE module

In IDRISI, another procedure, Ordered Weighted Average (OWA), is provided in MCE module. Ordered Weighted Average procedure will be discussed in Chapter 2. For now, we will discuss the calculation method of criteria weights firstly.

1.2 Criteria Weights Calculation

If there are more than 3 criteria, weights calculation becomes quite difficult. The technique implemented in IDRISI for criteria weights calculation is Analytical Hierarchy Process (AHP) method. In AHP, criteria weights are obtained from the principal eigenvector of a square reciprocal matrix of pairwise comparisons. The relative importance of the two criteria is used to create pairwise comparisons. For example, a 9-point continuous scale is shown as Figure 1-3. In this scale, 5 means one factor was strongly more important than another factor, on the inverse case, it would be 1/5.

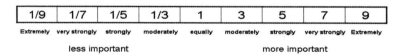

Fig.1-3　Continuous rating scale

In this method, criteria weights are an eigenvector, for maximum latent root λ_{max} in comparison matrix A, eigenvector is W, the calculation of criteria weights is to calculate eigenvector W, makes,

$$AW = \lambda_{max} W \tag{1-4}$$

Firstly,

$$\bar{a}_{ij} = \frac{a_{ij}}{\sum_{k=1}^{n} a_{kj}} \quad i,j = 1,2,\ldots,n \tag{1-5}$$

Then to calculate vector as,

$$\overline{W}_i = \sum_{j=1}^{n} \bar{a}_{ij} \quad i,j = 1,2,\ldots,n \tag{1-6}$$

To standardize vector as,

$$W_i = \frac{\overline{W}_i}{\sum_{j=1}^{n} \overline{W}_j} \quad i,j = 1,2,3,\ldots,n \tag{1-7}$$

The criteria weights are,

$$u_i = W = [W_1, W_2, \ldots, W_n]^T \tag{1-8}$$

Maximum latent root is calculated as follows,

$$\lambda_{max} = \sum_{i=1}^{n} \frac{(AW)_i}{nW_i} \tag{1-9}$$

In which,

$(AW)_i$ = the i-the element in AW.

The calculation of C.I. is,

$$\text{C.I.} = \frac{\lambda_{max} - n}{n - 1} \tag{1-10}$$

Randomly consistency index (R.I.) is offered in Table 1, C.R. is calculated as,

$$\text{C.R.} = \frac{\text{C.I.}}{\text{R.I.}} \tag{1-11}$$

If the value of consistency ratio from Formula (1-11) is less than 0.10, it indicates a good consistency, and criteria weights are acceptable. If the value exceeds 0.10, the comparison matrix needs to be modified, and criteria weights should be re-calculated. In IDRISI, this can be fulfilled in WEIGHT module.

Table 1-1   Randomly consistency index

matrix order	1	2	3	4	5	6	7	8
R.I.	0	0	0.58	0.90	1.12	1.24	1.32	1.41
matrix order	9	10	11	12	13	14	15	
R.I.	1.46	1.49	1.52	1.54	1.56	1.58	1.59	

In addition to the overall analysis, it is necessary to analyze the matrix to analyze where the inconsistencies arise.

As a simple application example, one comparison matrix with only 3 factors is constructed as,

$$A = \begin{bmatrix} 1 & 2 & 6 \\ 1/2 & 1 & 4 \\ 1/6 & 1/4 & 1 \end{bmatrix} \quad (1-12)$$

The weights can be achieved by calculating the weights with each column and then averaging over all columns. For example, if we take the first column of Figures, they sum to 1.67. Dividing each of the entries in the first column by 1.67, it yields weights of 0.60, 0.30, and 0.10. Repeating this for each column and averaging the weights over the columns usually gives a good approximation to the values calculated by the principal eigenvector. The results of Formula (1-12) is,

$$w = \begin{bmatrix} 0.588 & 0.322 & 0.090 \end{bmatrix}^T \quad \lambda = 3.010 \quad (1-13)$$

Because $\lambda = 3.010 \neq 3$, it shows the reason of inconsistency, and it comes from the comparison matrix.

We modify Formula (1–12),

$$A = \begin{bmatrix} 1 & 2 & 6 \\ 1/2 & 1 & 3 \\ 1/6 & 1/3 & 1 \end{bmatrix} \quad (1-14)$$

The results of Formula (1–14) is,

$$w = \begin{bmatrix} 0.6 & 0.3 & 0.1 \end{bmatrix}^T \quad \lambda = 3 \quad (1-15)$$

Through the comparing of Formula (1–12) with Formula (1–14), it is found that the lack of consensus comes from the inconformity of $A1$, $A2$, and $A3$. For example, if $A1/A2=2$, $A1/A3=6$, then should have the equation: $A2/A3=3$. But in Formula (1–12), $A2/A3=3.010 \neq 3$, this is the reason of lack of consensus. On the other hand, in Formula (1–14), $A2/A3=3$, so it is complete consensus.

In fact, many application examples have more than 3 factors, such as the pairwise comparison matrix in Table 1–2. Since the matrix is symmetrical, only the lower triangular half actually needs to be filled in. The remaining cells are then simply the reciprocals of the lower triangular half (for example, since the rating of $B3$ to $B1$ proximity is 4, the rating of $B1$ proximity relative to $B3$ will be 1/4).

Table 1-2 Comparison matrix of four factors

Factor	B1	B2	B3	B4
B1	1			
B2	1/2	1		
B3	4	6	1	
B4	1/5	1/3	1/8	1

The calculating results of criteria weights and consistency ratio are,

$$w = \begin{bmatrix} 0.2077 & 0.1164 & 0.6248 & 0.0511 \end{bmatrix}^T \quad \text{C.R.} = 0.04 < 0.1 \quad (1-16)$$

According to Table 1–2, we construct an improved comparison matrix as Table 1–3. Corresponding to the values of Table 1–2, we can find the effect of altering any of the pairwise comparisons. And the key to consensus is to keep the ratio inconformity of all factors.

Table 1–3 Improved comparison matrix of four facters

Factor	B1	B2	B3	B4
B1	1			
B2	1/2	1		
B3	4	8	1	
B4	1/5	2/5	1/20	1

The calculating results are,

$$w = [0.1754 \quad 0.0877 \quad 0.7018 \quad 0.0351]^T \quad \lambda = 4 \quad C.R. = 0 \quad (1-17)$$

The higher the weight, the more important the factor in determining suitability for the objective will be.

1.3 Application Examples

Three application examples are introduced as follows.

1.3.1 An application example for Boolean Intersection

Problem description: Tangshan is city with frequently disasters. For example, the big earthquake in 1976 deprived more than 240 000 people's lives. Besides earthquake, Tangshan City is threatened by many disasters, such as karst collapse, goaf collapse, and so on. Thus, more attention is paid to site safety evaluation, and disaster images have been exported as MIF files from MapInfo software.

In IDRISI, from File menu, select Import ... Software-specific For-

Chapter 1 Multi-Criteria Evaluation

mats ... MIFIDRIS (MapInfo), see Figure 1-4.

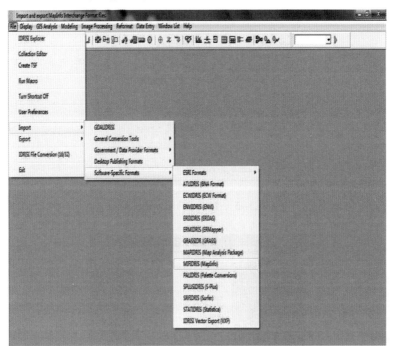

Fig.1-4   File input operation

In the pop up windows, choose files for Input MMIF file, Output Idrisi vector file. Choose "plane" for Reference system, "meters" for Reference units, and "1.0" for Unit distance. The Features to be processed is "regions", and input proper title for your operation. See Figure 1-5.

Then, choose "Raster / Vector Conversion" in menu Reformat (Figure 1-6), the raster images were obtained. See Figure 1-7 to Figure 1-10.

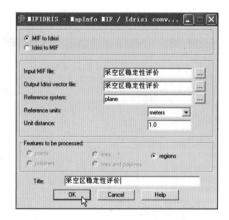

Fig.1-5　MIFIDRIS window

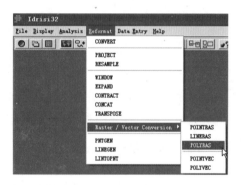

Fig.1-6　Vector to raster

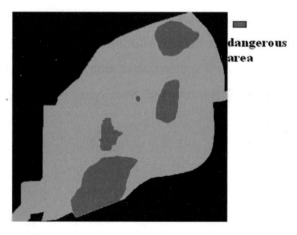

Fig.1-7　Dangerous area for goaf collapse

Chapter 1 Multi-Criteria Evaluation

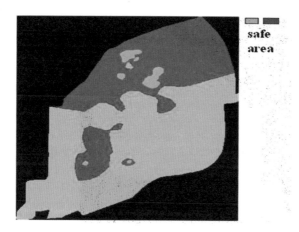

Fig.1-8 Safe area for karst collapse

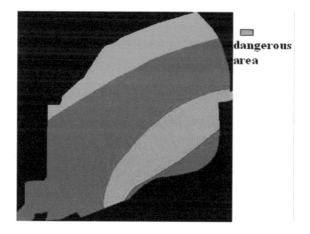

Fig.1-9 Dangerous area for earthquake

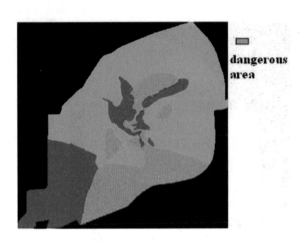

Fig.1-10 Dangerous area for site condition

Figure 1-7 to Figure 1-10 show the locations of factors for site safety evaluation.

There are four factors, and those raster images are dangerous areas of goaf collapse, earthquake, site condition and safe area of karst collapse. Obviously, for dangerous area, there is higher suitability in land use far away dangerous area. Therefore, the safety suitability depends up the distance for each factor from dangerous area or safe area.

Through RECLASS operation or image calculator, Boolean images with only values of 1 or 0 are obtained as Figure 1-11 to Figure 1-14.

Figure 1-11 is the picture of goaf collapse. In this picture, we know dangerous areas. Because goaf collapse areas are not suitable for building, therefore, value 0 is given for goaf collapse areas, and other areas with value 1. Figure 1-12 is the picture of karst collapse, here, we know the safe areas, and value 1 is assigned to those areas.

Chapter 1　Multi-Criteria Evaluation

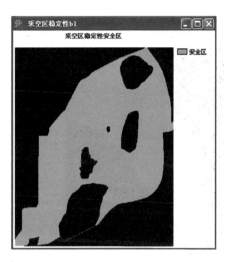

Fig.1-11　Boolean image of goaf collapse safe areas

Fig.1-12　Boolean image of karst collapse safe areas

Figure 1-13 is the picture of earthquake influence. In this picture, we know very dangerous areas, and the NOT very dangerous areas are assigned with value 1. Figure 1-14 is the picture of site condition, in this picture, good site condition areas are assigned with value 1, and other areas are given value 0.

Fig.1-13   Boolean image of earthquake safe area

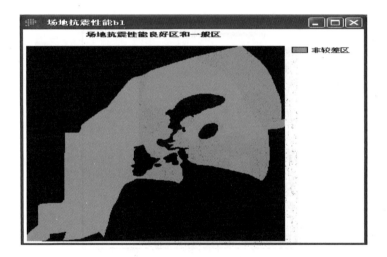

Fig.1-14   Boolean image of site condition safe area

Now, we evaluate the site safety according to above four factors with Boolean Intersection. From the GIS Analysis menu, we choose Decision Support ... MCE, see Figure 1-15.

Chapter 1 Multi-Criteria Evaluation

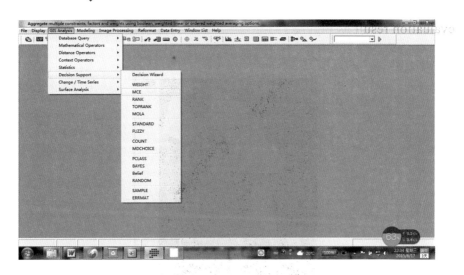

Fig.1-15 MCE operation window

In pop up MCE window, choose Boolean intersection, see Figure 1-16.

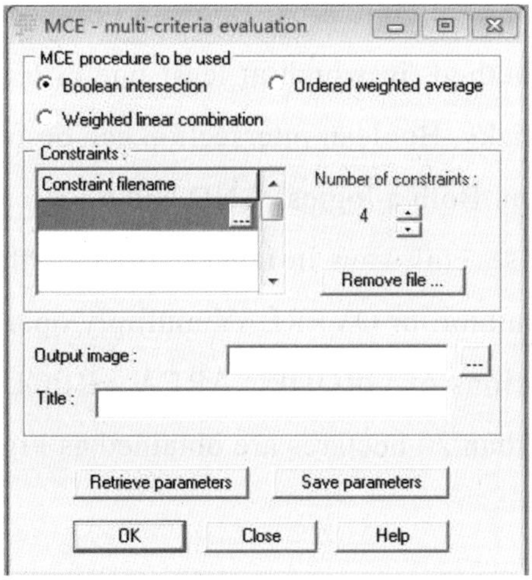

Fig.1-16 Boolean intersection, window

Reference Figure 1-16, Number of constraints is assigned as 4, then, choose four constraint files (influence factors), they are files of Figure 1-11 to Figure 1-14. Given output image name and title, and click OK to obtain evaluation result as Figure 1-17.

Fig.1-17 Suitable area of BI

Boolean intersection is the operation of logical AND, that requires all criteria to be met for a suitable area. Boolean union or logical OR is another aggregation method, in which at least one criteria is requires to be met, thus it is very risky. Boolean intersection can be fulfilled in MCE, but on for Boolean union. Both a logical AND and a logical OR operation can be accomplished in several ways in IDRISI, such as the Decision Making Wizard, Image Calculator, or OVERLAY multiply operations.

Through operations of GROUP、AREA、RECLASS、OVERLAY, suitable areas large than 20 hectares are obtained as Figure 1-18.

Chapter 1　Multi-Criteria Evaluation

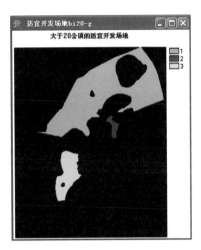

Fig.1-18　Suitable areas (>20 hectares)

In Figure 1-18, there are three suitable areas large than 20 hectares for building construction. They are label as site 1, 2, and 3, site 1 with area of 6871.85 hectares, site 2 with 204.441 hectares, and site 3 with 1233.498 hectares.

In Boolean intersection, one criterion cannot compensate suitability from any other. In other word, only by exactly meeting all criteria is a location considered suitable. The result is the best location possible for residential development and no less suitable locations are identified.

1.3.2 An application example for WLC

In this exercise, land use suitability is no longer simply divided into suitability and unsuitability. Factors will be standardized to continuous scales. We know that suitability is related to distance, the far from dangerous areas, and the higher in land use suitability. Therefore, the distance from dangerous or safety areas are calculated.

According to the images from Figure 1-7 to Figure 1-10, Module DISTANCE in IDRISI is used to calculate the distance from the most dangerous (or safety) area. The dangerous areas in Figure 1-7, Figure 1-8, and Figure 1-10, are assigned as feature image, see Figure 1-19. But for Figure 1-8, the safety areas are assigned as feature image.

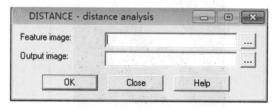

Fig.1-19　Distance operation

The distance images are shown as Figure 1-20 to Figure 1-23.

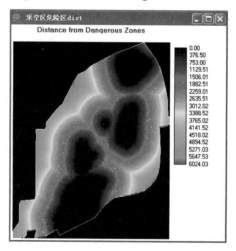

Fig.1-20　Distance for goaf collapse

Chapter 1 Multi-Criteria Evaluation

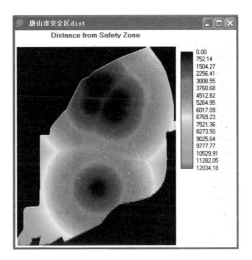

Fig.1-21　Distance for karst collapse

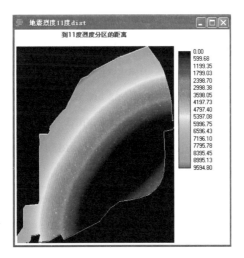

Fig.1-22　Distance for earthquake

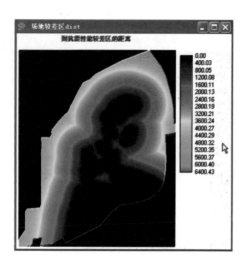

Fig.1-23　Distance for site condition

Now, we standardize all of those factors to continuous scales.

In order to evaluate land use suitability under standard values, all factors' images are standardized in terms of a certain criteria with FUZZY module in IDRISI. Fuzzy set membership functions are offered in IDRISI to standardize a factor by FUZZY module is shown in Figure 1-1, in which three kinds of optional function are included, such as Sigmoid, J-shaped and Linear. The first disaster factor is goaf collapse. According to the structure of overlaying rock and soil, influencing distance of goaf collapse is 596,632,668, and 705 meters. The far most distance is 705 meters. Because the far from goaf collapse area, the higher in site safety, ascend sigmoid function is selected as evaluation criterion, with point 'a' equals to 0, point 'b, c, and d' equals to 750, see Figure 1-24(a).

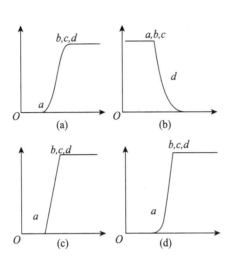

Fig.1-24   Evaluation criteria

The second disaster factor is karst collapse. For karst collapse, it is easy to know the position of safety area. The far from safety area, the lower site safety is. Such, descend J-shaped function is selected for karst collapse, with point 'a, b, and c' equals to 0, point 'd' equals to 12035, see Figure 1-24(b). The third disaster factor is earthquake, and the most dangerous area is highest earthquake intensity area in history. The far most distance from highest earthquake intensity area is 9500 meters, where the earthquake intensity is half of the highest area. Thus, ascend sigmoid function is selected for earthquake, with point 'a' equals to 0, point 'b, c, and d' equals to 19000, see Figure 1-24(c). The fourth disaster factor is site condition, by surveying the distance from the unfavorably earthquake resistant area, ascend J-shaped function is selected, with point 'a' equals to 1000, point 'b, c, and d' equals to 6000, see Figure 1-24(d).

The FUZZY module can be started from decision support of menu "GIS analysis" in IDRISI, see Figure 1-25. And the FUZZY window is shown in Figure 1-26.

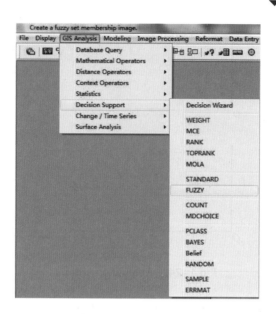

Fig.1-25   FUZZY module in IDRISI

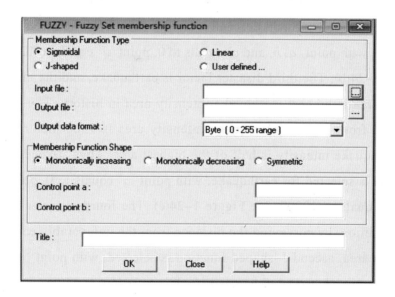

Fig.1-26   FUZZY operation window

FUZZY images of four factors are shown from Figure 1-27 to Figure 1-30.

Chapter 1　Multi-Criteria Evaluation

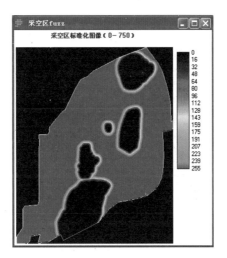

Fig.1-27　FUZZY image of goaf collapse

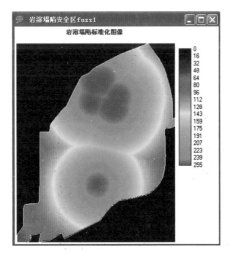

Fig.1-28　FUZZY image of karst collapse

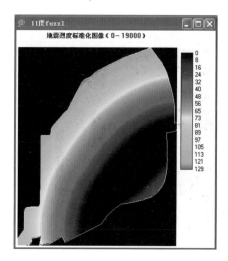

Fig.1-29　FUZZY image of earthquake

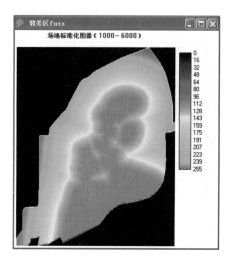

Fig.1-30　FUZZY image of site condition

It can be found that the highest suitability is 255 in Figure 1-27, 1-28, and 1-30, but only 129 in Figure 1-29. The reason is that the selection of criterion function in FUZZY module, see Figure 1-24(c), especially the value of point "b". It means that earthquake disaster can not be avoid in Tangshan City, but can be reduced.

Next, before evaluation, we calculate criteria weights firstly. WEIGHT module is also a module in decision support, see Figure 1-25. In WEIGHT module, corresponding comparison matrix can be constructed. Each value of comparison matrix reflects the relative importance of factors. According to the AHP method that has been introduced in 1.2 of this chapter, comparison matrix is shown as Table 1-4.

Table 1-4 Comparison matrix

Disaster factor	Goaf collapse	Karst collapse	Earthquake	Site condition
Goaf collapse	1	2	1/4	5
Karst collapse	1/2	1	1/6	3
Earthquake	4	6	1	8
Site condition	1/5	1/3	1/8	1

If comparison matrix is constructed in WEIGHT module, the number of factors and the file name of standardized images must be specified. Of course, a comparison matrix can also be edited in EDIT module to make an attribute file, and then use it directly in WEIGHT module. The comparison matrix produced by this technique, is a positive reciprocal matrix. Therefore, only the higher/lower triangular half which includes $n(n-1)/2$ elements needs to be filled in.

The results of criterion weights for goaf collapse, karst collapse, earthquake, and site condition are 0.2077, 0.1164, 0.6248, and 0.0511. Consistency Ratio is 0.04, it is less than 0.10, and indicate good consistency. If values of C.R. exceed 0.10, the comparison matrix needs to be modified, and the matrix of weightings should be re-evaluated.

Finaly, choose MCE module in decision support of GIS analysis menu, and pick WLC in MCE window, see Figure 1-31.

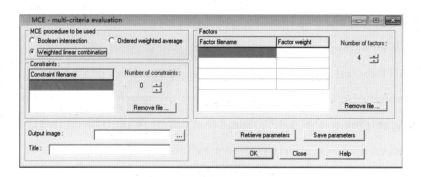

Fig.1-31　MCE window

Choose Number of factors as 4, Number of constraints as 0 (you can guess that sometimes we need one more constraints). Then make choice of factor filename and factor weights for those four factors, of course, output image and title also should be inputted. Click OK for WLC evaluation. The evaluation result is shown in Figure 1-32.

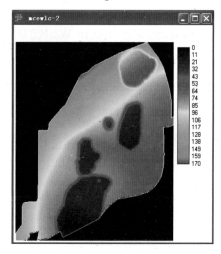

Fig.1-32　WLC result of land use suitability

Figure 1-32 is the land use suitability affected by disaster factors in Tangshan City. From Figure 1-32, it can be found that some unsuitable areas distributed in center district, so more attention should be paid for those areas and lower density for construction should be insisted. Also, the suitability for land use increases gradually from southeast to northwest. Therefore, city construction should be extended to the northwest in city planning.

From above application, we can found that IDRISI is a kind of GIS analysis software with powerful capability for decision analysis, and it is applied to the decision analysis of land use based on disaster factors successful. The right solution is solved through the constructing of comparison matrix with AHP method and weights matrix is calculated. This process can also apply to other fields of decision analysis, and IDRISI should be applied extensively in the future.

Once the weights matrix has been determined, the module MCE can be used to aggregate factor images and the suitability image for Multi-Criteria Evaluation will be obtained. WLC is one of the three decision support tools for Multi-Criteria Evaluation. Under general condition, WLC is used for suitability evaluation. The aggregation method of Weighted Linear Combination (WLC) multiplies each standardized factor image by its factor weight and then sums the results. This result is then multiplied by each of the constraints in turn to "mask out" unsuitable areas.

IDRISI also includes a third option for Multi-Criteria Evaluation, known as an Ordered Weighted Average (OWA). This method offers a complete range of decision strategies along the degree of Tradeoff involved and degree of risk in the solution. It will be discussed in Chapter 2 of decision support in this book.

Chapter 2　Ordered Weighted Averaging

There are two fundamental classes of multi-criteria evaluation in GIS: one is the Boolean overlay operation based on Boolean calculation of intersection (AND) and union (OR), and the other is weighted combination based on evaluation criteria with sequential standardization, which includes classic Weighted Linear Combination (WLC) method and Ordered Weighted Averaging (OWA) method (Malczewski, 2006). Ordered Weighted Averaging (OWA) is a relatively new method was introduced by Yager in 1988, and quantifier guided aggregation was given in 1996. Ordered weighted averaging is a multi-criteria evaluation procedure, which can generate a wide range of decision strategies through calculating order weights and criteria weights.

Salem Chakhar et al. (2003) discussed the combination between GIS and multi-criteria evaluation, and the enhancing capabilities for GIS with multi-criteria evaluation functions. Jacek. Malczewski (2004, 2006) analyzed GIS-based multi-criteria evaluation for land-use suitability and the application of OWA method to identify the most suitable lands for housing development. Because of the uncertainty of influencing factors and attribute values, and standardization with different criteria, Qiu Bingwen et al. (2004) indicated that the main problems of evaluation methods were factors selection and standardization, criteria weights calculation and the combination of GIS with decision-making procedures. As basis of many type statistical data, spatial model for city land use was constructed by Li Zhenguo (2005), and was applied to spatial decision-making for land use.

About OWA method, many results have been obtained for calculation of OWA operators, and applied to many domains without GIS environment, such as business evaluation, multi-attributes decision-making, and image analysis, and so on.

Boolean calculation of intersection is evaluated as the minimum value for a pixel across all factors corresponding to the MIN operator. Boolean calculation of union means that at least one is satisfied, which is evaluated as the maximum value for a pixel across all factors corresponding to the MAX operator. Obviously, no Tradeoff or compensation is considered in Boolean overlay. In weighted combination methods, a factor with a high criterion weight can Tradeoff or compensate for poor weights on other factors. WLC method is situated at the mid-point on the continuum ranging from the MIN (Boolean 'AND' operator) to MAX (Boolean 'OR' operator), which indicates full Tradeoff among criteria; OWA method can select any degree of Tradeoff among criteria between no Tradeoff and full Tradeoff according to the decision-making strategy. Therefore, Boolean overlay represents the extreme cases with no Tradeoff, Boolean 'AND' operator represents the MIN risk decision making, and Boolean 'OR' operator represents the MAX risk decision-making in strategy. WLC method is an averaging risk decision-making with full Tradeoff among criteria; OWA method can select risk dynamically according to decision-making strategy, and obtain any results from the MIN risk to MAX risk with appropriate Tradeoff. In site safety evaluation, not only the criterion weights but also order weights need to be calculated for each factor. In different disaster prevention strategy, order weights need to be calculated according to the important rank of each disaster factor. Therefore, OWA method is very important for site safety evaluation in city land use planning.

2.1 Method Introduction

Different with Formula (1-1), in GIS-based OWA method, both criteria weights and order weights are considered. For one spatial location or pixel (x,y), v_i is the i-th order weight, and the calculation of suitability is expressed as,

$$f(x,y) = \sum_{i=1}^{m} \left(\frac{u_i v_i}{\sum_{i=1}^{m} u_i v_i} \right) z_i(x,y) \qquad (2\text{-}1)$$

The weights of OWA are defined as,

$$w_i = \frac{u_i v_i}{\sum_{i=1}^{n} u_i v_i} \qquad (2\text{-}2)$$

A measure of Orness and Tradeoff, associated with a particular set of weights can be obtained by the following equations (Yager, 1988).

$$\text{Orness} = \alpha = \sum_{i=1}^{n} \frac{n-i}{n-1} v_i \qquad (2\text{-}3)$$

In which, $\alpha \in [0,1]$.

$$\text{Tradeoff} = 1 - \sqrt{\frac{n \sum_{i=1}^{n} (v_i - \frac{1}{n})^2}{n-1}} \qquad (2\text{-}4)$$

If the set of weights are w_j (OWA weights), then Orness and Tradeoff for OWA method can be expressed as,

$$\text{Orness}(w) = \alpha = \sum_{i=1}^{n} \frac{n-i}{n-1} w_i \qquad (2\text{-}5)$$

$$\text{Tradeoff}(w) = 1 - \sqrt{\frac{n \sum_{i=1}^{n} (w_i - \frac{1}{n})^2}{n-1}} \qquad (2\text{-}6)$$

If Boolean constraints are considered, the Formula (2-1) can be changed as,

$$f(x,y) = \sum_{i=1}^{m} \frac{u_i v_i}{\sum_{i=1}^{m} u_i v_i} z_i(x,y) \times \prod c_j(x,y) \qquad (2-7)$$

In which,

c_j = criterion score of constraint j at point (x, y),

$\prod$ = product.

Orness can be regards as risk, the relation between risk and Tradeoff can be shown as Figure 2–1.

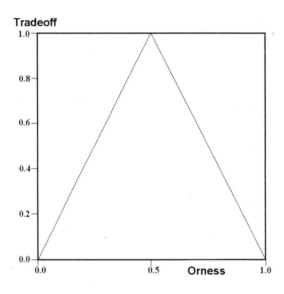

Fig. 2–1 Relation between risk and Tradeoff

In Figure 2–1, Boolean intersection is evaluated as the minimum risk 0. Boolean union means the risk. Obviously, no Tradeoff or compensation is considered in Boolean overlay. In Weighted Linear Combination (WLC) methods, a factor with a high criterion weight can Tradeoff or compensate for poor weights on other factors. WLC method is situated at 0.5 on the continuum ranging from the MIN to MAX risk, which indicates full Tradeoff among criteria. OWA method can select any degree of Tradeoff among criteria between no Tradeoff and full Tradeoff according to the decision-making strategy. OWA method can also select risk according to decision-making strategy, and obtain any results from the MIN risk to MAX risk with appropriate Tradeoff.

In order to calculate suitability, calculation of both order weights and criteria weights are required, which is discussed in 2.2.

2.2 Order Weights Calculation

The first and simplest calculation method is with rank. Ordering weights v_k is to be determined by their rank position r_k, v_k is expressed as follows,

$$v_k = \frac{n - r_k + 1}{\sum_{l=1}^{k}(n - r_l + 1)} \quad (k = 1, 2, \ldots, n) \tag{2-8}$$

If there are four criteria, according to Formula (2-8), $r_1=1$, $r_2=2$, $r_3=3$, $r_4=4$, v_k is obtained, and the results are 0.4, 0.3, 0.2, and 0.1. Then, Orness and Tradeoff are calculated according to Formula (2-3) and Formula (2-4):

ORNESS(1)=0.4+2 × 0.3/3+0.2/3=0.6667

TRADEOFF(1)=0.7418

If $r_1=4$, $r_2=3$, $r_3=1$, $r_4=1$, then, v_k is calculated, and the results are 0.1, 0.2, 0.3, 0.4. According to Formula (2-3) and Formula (2-4):

ORNESS(2)=0.1+2 × 0.2/3+0.3/3=0.3333

TRADEOFF(2)=0.7418

The relationship between Orness and Tradeoff with order weighting vector $v=\{0.4, 0.3, 0.2, 0.1\}^T$ is shown as Figure 2-2.

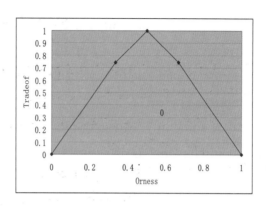

Fig. 2-2　Relation between risk and Tradeoff with rank

In Figure 2-2, OWA method offered more selections for decision making with different risk. In fact, order weighting vector $v=\{0.4, 0.3, 0.2, 0.1\}^T$ offered a selection with risk equal to 0.666 7 and the Tradeoff is 0.741 8. Corresponding, order weighting vector $v=\{0.1, 0.2, 0.3, 0.4\}^T$ offered a selection with risk equal to 0.333 3 and the Tradeoff is also 0.741 8.

There are several other methods to calculate order weights; the modified rank method is defined as follows,

$$v_j = \left(\sum_{k=1}^{j} v_k\right)^\beta - \left(\sum_{k=1}^{j-1} v_k\right)^\beta \tag{2-9}$$

In which, β is a parameter in range from 0 to ∞, and corresponds to the risk of decision making.

The difference in strategy of this method can be explained by a simple example application. For example, there are three evaluation factors A, B, and C, with attribute values $A=87$, $B=74$, and $C=101$. Criterion weights are 0.5, 0.2, 0.3. The sequence in decreasing order is $C>A>B$, and reordering criterion weights are 0.3, 0.5, and 0.2. In order to make the calculation easy, let r_k ($k=1,2,3$)=1/3, it is means that the importance of rank ordering is equal, the results are shown in Table 2-1.

Table 2-1 Comparing with different Orness

No.	Order weights			Criterion weights (u_1, u_2, u_3)	Calculating weights			Results	Methods	Weighting vectors	β	Orness
	v_1	v_2	v_3		w_1	w_2	w_3					
1	1.0	0.0	0.0	0.3, 0.5, 0.2	1.0	0.0	0.0	101	Boolean OR	$v_1=1, v_j=0$	$\beta \to 0$	1.0
2	0.58	0.24	0.18	0.3, 0.5, 0.2	0.53	0.36	0.11	92.99	OWA	v_1	$\beta=0.5$	1.0–0.5
3	0.34	0.33	0.33	0.3, 0.5, 0.2	0.3	0.5	0.2	88.6	WLC	$v_j=1/n$	$\beta=1$	0.5
4	0.12	0.33	0.55	0.3, 0.5, 0.2	0.12	0.53	0.35	84.13	OWA	v_1	$\beta=2$	0.5–0.0
5	0.0	0.0	1.0	0.3, 0.5, 0.2	0.0	0.0	1.0	74	Boolean AND	$v_n=1, v_j=0$	$\beta \to \infty$	0.0

In Table 2-1, $v_1=1$, and $v_j=0$ for all other weights corresponds to Boolean OR operator (No. 1 in Table 2-1), and the result is equal to the situation $\beta \rightarrow 0$ in Formula (2-9), represents maximum risk level. $v_n=1$, and $v_j=0$ for all other weights, corresponds to Boolean AND operator (No. 5 in Table 2-1), and is associated with the situation $\beta \rightarrow \infty$ in Formula (2-9), represents minimum risk level. Assigning equal order weights for all evaluation criteria ($v_j=1/n, j=1, 2, \ldots, n$) results in the conventional WLC method, corresponds to the situation $\beta=1$ in Formula (2-9), represents average risk level. Consequently, both Boolean overlay and WLC represent the special cases of OWA, and correspond to special values of β in Formula (2-9) with special risk level. OWA method can select risk level according to decision-making strategy. For example, $\beta=0.5$, and $\beta=2$, correspond to the risk level of a few evaluation criteria and most evaluation criteria are satisfied respectively (No.2 and No.4 in Table 2-1), appropriate results can be obtained by OWA method according to the decision-making strategy.

The difference in decision-making strategy for different methods can be explained through weighting vector $v=\{v_1, v_2, \ldots, v_n\}^T$, $v=\{0, 0, \ldots, 1\}^T$ in OWA corresponds to Boolean AND operator, $v=\{1, 0, \ldots, 0\}^T$ in OWA corresponds to Boolean OR operator. $v=\{1/n, 1/n, \ldots, 1/n\}^T$ in OWA corresponds to WLC, represents average risk level. Consequently, OWA can generate a wide range of decision strategies with weighting vector changing from $v=\{0, 0, \ldots, 1\}^T$ to $v=\{1, 0, \ldots, 0\}^T$ continuously. Therefore, applying GIS-based OWA method to multi-criteria evaluation in geographical spatial analysis, can obtain exact evaluation results with any decision risk level.

The Orness and Tradeoff according to Formula (2-9) is shown in Table 2-2 with rank weights (0.4, 0.3, 0.2, 0.1).

Table 2-2 Orness and Tradeoff with different β

β	0.25	0.5	0.8	1.2	1.5	2
Orness	0.894 7	0.806 0	0.717 2	0.622	0.564 2	0.486 7
Tradeoff	0.268 8	0.474 7	0.655 5	0.802 5	0.845 3	0.824 9

The relation between β and Orness is shown in Figure 2-3.

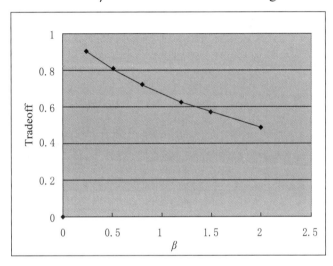

Fig. 2-3 Relation between β and Orness

From Figure 2-3, it can be found that Orness decreases gradually with the increase of Orness. The difference is that the relation between risk and Tradeoff is not linear but a curve.

The relation between β and Tradeoff is shown in Figure 2-4. In Figure 2-4, Tradeoff increases gradually with the increase of Orness. But Tradeoff can not reach 1, it means that full compensation can not achieved. Therefore, the curve shape and maximum compensation point value depend on β.

The relation between Orness and Tradeoff with this method is shown in Figure 2-5. In Figure 2-5, β can represents risk.

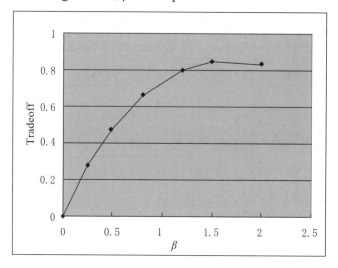

Fig. 2-4 Relation between β and Tradeoff

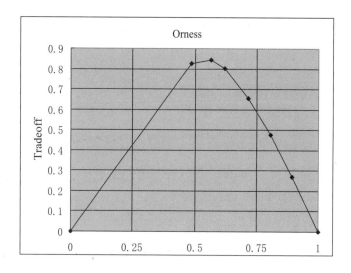

Fig. 2-5 Relation between Orness and Tradeoff with β

When rank is equal with (0.25,0.25,0.25,0.25), the value of Orness and Tradeoff are shown in Table 2-3.

Chapter 2 Ordered Weighted Averaging

Table 2-3 Orness and Tradeoff with equal rank

β	0.5	0.8	1	1.5	2
Orness	0.691 0	0.566 2	0.5	0.376	0.291 7
Tradeoff	0.661 2	0.888 7	1	0.805 7	0.677 2

With this condition, the relation between Orness and Tradeoff with this method is shown in Figure 2-6.

From all above Figures, it can be found that β can represents risk or Orness, only with different scales. The risk of decision making can be chose through selection of value of β. The value of rank dictates the curve shape between Orness and Tradeoff, also the maximum compensation point value.

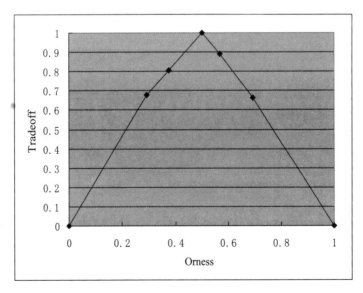

Fig. 2-6 Relation between Orness and Tradeoff with β

Xu Zeshui et al.(2003) provides one method for the calculation of order weights with given Orness. For Orness= α, randomly generates $n+1$ nonnegative real number p_i, $p_i - p_{i-1} > 0$, ($i=1,2,...,n$), and $p_0=0$. Then, calculate q_i and α', as follows,

$$q_i = \sum_{j=1}^{i} p_i, s_i = q_i / q_n, \alpha' = \sum_{i=1}^{n-1} s_i / (n-1)(i = 0,1,\ldots,n) \quad (2\text{--}10)$$

If $\alpha' > \alpha$, go to Formulas (2–11) and (2–12), order weights v_i can be calculated. Otherwise, go to Formulas (2–13) and (2–14) to calculate order weights.

$$\alpha = \alpha / \alpha', \quad s'_n = s_n, \quad s'_i = s_i c (i = 0, 2, \ldots, n-1) \quad (2\text{--}11)$$

$$v_i = s'_i - s'_{i-1} \quad (2\text{--}12)$$

$$s'_i = s_i + it, \quad \sum_{i=1}^{n-1} s'_i / s_n / (n-1) = \alpha \quad (2\text{--}13)$$

$$v_i = (s'_i - s'_{i-1}) / s'_n \quad (2\text{--}14)$$

In this method, with Formula (2–10) to Formula (2–14), the relationship between Orness and Tradeoff is shown as Figure 2–7. It is found that the Tradeoff is 0.80 with Orness "0.75", it means that this method offer a much larger Tradeoff at the same Orness point of order weighting $v=\{0.4, 0.3, 0.2, 0.1\}^T$ or vector $v=\{0.1, 0.2, 0.3, 0.4\}^T$.

Unfortunately, in this method, any Orness is corresponding to a fixed Tradeoff. It means that only one fixed Tradeoff can be selected with a selected Orness, and no other selection is provided. All in all, evaluation results with any decision risk level can be obtained by this method, although it is fixed with Tradeoff.

Chapter 2 Ordered Weighted Averaging

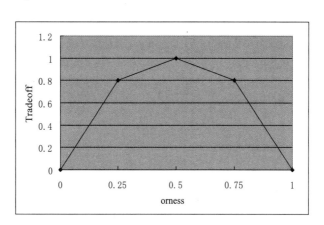

Fig. 2-7 Relation between Orness and Tradeoff with α

The difference in decision making strategy among different methods can be explained by a simple example application. With attribute values $A=87$, $B=74$, and $C=101$ of three factors. Criterion weights are 0.5, 0.2, and 0.3. The sequence in decreasing order is $C>A>B$, and reordering criterion weights are 0.3, 0.5, and 0.2. According to Formula (2-10) to Formula (2-14), the results are shown in Table 2-4.

Table 2-4　Results with different decision-making methods

No.	Order weights			Criteria weights (u_1, u_2, u_3)	OWA weights			Results	Corresponding methods	Weights analysis	Orness	Tradeoff
	v_1	v_2	v_3		w_1	w_2	w_3					
1	0.0	0.0	1.0	0.3, 0.5, 0.2	0.0	0.0	1.0	74	Boolean AND	$v_n=1, v_j=0$	0	0
2	0.14	0.22	0.64	0.3, 0.5, 0.2	0.09	0.33	0.58	80.7	OWA	$v_1, v_2, \cdots, v_n$	0.25	0.535
3	1/3	1/3	1/3	0.3, 0.5, 0.2	0.3	0.5	0.2	88.6	WLC	$v_j=1/n$	0.5	1
4	0.64	0.22	0.14	0.3, 0.5, 0.2	0.58	0.33	0.09	94.0	OWA	$v_1, v_2, \cdots, v_n$	0.75	0.535
5	1.0	0.0	0.0	0.3, 0.5, 0.2	1.0	0.0	0.0	101	Boolean OR	$v_1=1, v_j=0$	1	0

Chapter 2　Ordered Weighted Averaging

OWA method can select risk level according to decision-making strategy. For example, Orness=0.25, and Orness=0.75, correspond to the risk level of most evaluation criteria and a few evaluation criteria are satisfied respectively (No.2 and No.4 in Table 2-4), appropriate results can be obtained by OWA method according to the decision-making strategy.

Influence of criteria weights on the calculating results can be analyzed through comparing the results between Orness and Orness(w). The results are shown in Table 2-5. In Table 2-5, 'Results' represents the results with only order weights, and 'Results (w)' represents the results with both order weights and criteria weights. In fact, the calculating results are defined by risk. When risk increase, the result (or suitability) increase correspondingly. Also, criteria weights make the Tradeoff among factors changed.

Table 2-5 Comparing between Orness and Orness(w)

No.	Order weights			OWA weights			Results	Results(w)	Orness	Orness(w)	Tradeoff	Tradeoff(w)
	v_1	v_2	v_3	w_1	w_2	w_3						
1	0.0	0.0	1.0	0.0	0.0	1.0	74.0	74.0	0	0	0	0
2	0.14	0.22	0.64	0.09	0.33	0.58	80.6	80.7	0.25	0.255	0.535	0.576
3	1/3	1/3	1/3	0.3	0.5	0.2	87.3	88.6	0.5	0.55	1	0.507
4	0.64	0.22	0.14	0.58	0.33	0.09	94.1	94.0	0.75	0.745	0.535	0.576
5	1.0	0.0	0.0	1.0	0.0	0.0	101.0	101.0	1	1	0	0

Therefore, with Formula (2–10) to Formula (2–14), a new method is offered for the calculation of order weights with given Orness.

2.3 Application Examples

2.3.1 Application example 1

In 1.3.2, an application example for WLC is introduced. Comparing with Figure 1–31, the OWA operation is shown in Figure 2–8.

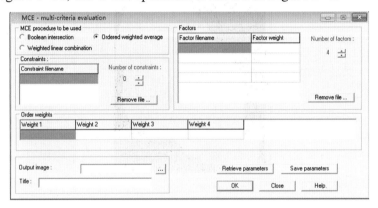

Fig. 2–8 OWA operations in MCE

It can be found in Figure 2–8, the difference between WLC and OWA operation is Order weights option. Therefore, criterion weights for goaf collapse, karst collapse, earthquake, and site types are same with the application example in 1.3.2, the results are 0.2077, 0.1164, 0.6248, and 0.0511.

After the calculation of criterion weights, order weights are also calculated. According to the Formula (2–8), $r_1=1$, $r_2=2$, $r_3=3$, $r_4=4$, then, wk is calculated, and the results are 0.4, 0.3, 0.2, and 0.1. Let risk degree index $\beta=0.6$, according to Formula (2–9), order weights are calculated, and the results are 0.5771, 0.2303, 0.1314, and 0.0612.

Once order weights and criterion weights have been determined, by the application of OWA method in IDRISI (Figure 2-8), the suitability image of site safety is obtained as Figure 2-9. Compare this image with Figure 1-31, it can be found that maximum suitability value increases from 170 to 210. The reason is that low risk strategy is selected than average risk (Figure 1-31), and this also makes the favorable area increasing.

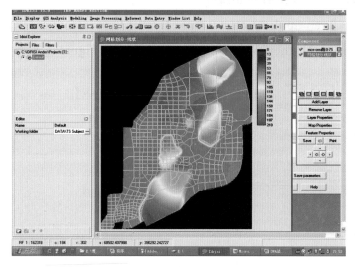

Fig.2-9   Suitability with OWA method

It is also found that the distribution trends of land use suitability in two Figures are similar, the suitability for land use increases gradually from southeast to northwest. Because high safety index is selected in each disaster factor evaluation, and some measures make the capability of building for disaster prevention increased in a few favorable areas in city construction, low risk strategy is rational. Therefore, more favorable areas for construction are obtained with appropriate risk selection.

According to the data in Tangshan City, there are 147 residential areas, floor area ratios of 42 residential areas are less than 0.8 (low density), 28.6 percentage of all residential areas. See Figure 2-10.

Chapter 2 Ordered Weighted Averaging

Fig.2-10　Floor area ratio of residential area

Comparing the distribution of floor area ratio with land use suitability, it is found that the distribution trend of floor area ratio is similar with land use suitability, which increases gradually from southeast to northwest. However, many residential areas distributed in southwest of center district, where suitability is low and more attention should be paid for those areas, lower density for construction should be insisted to avoid the lost of geological disaster. No residential areas are developed in the northwest of Tangshan City, where the land use suitability is high. Therefore, construction of residential area should be extended to the northwest to insure rational development of land resource.

Next, we will discuss the risk change with different β in Formula (2-9). For example, β =0.5, the order weights are (0.5, 0.207 1, 0.158 9, 0.134 0), and Orness is 0.691 0, with the maximum suitability value 203. From Formula (2-5) and Formula (2-6), we can calculate the total risk and Tradeoff (according to the OWA weights, not order weights). According to Formula

(2-2), the probably MIN and MAX OWA weights sets are (0.820 5, 0.113 0, 0.048 5, 0.018 0) with Orness 0.912, and (0.153 5, 0.144 9, 0.198 4, 0.503 2) with Orness 0.316. Therefore, for OWA evaluation, the risk or Orness is changed in different location (or spatial point or pixel). In this example, the risk is change from 0.316 to 0.912, not exactly 0.691. If the v_j is equal in Formula (2-9), the result image is shown as Figure 2-11.

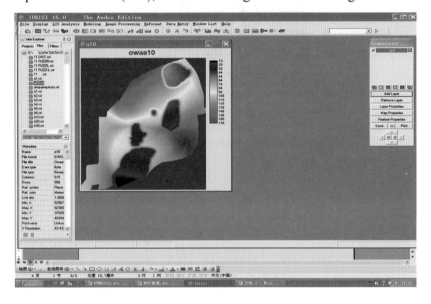

Fig.2-11　Result with equal rank in Formula 2-9

2.3.2 Application example for sample points data

Problem description: In J25 block, Jinzhou oilfield, there are 120 wells. The oil production is through thermal injection, vapors with high temperature and pressure are injected into strata through wells, but this result in casing failure. In order to protect well casing, value of vapor injection, injection number, and injection pressure are use to predict the risk of casing failure with WLC and OWA evaluation.

The data of well thermal injection is shown in Table 2-6.

Table 2-6 Well data of vapor injection

No.	X-axis	Y-axis	Amount of vapour	Injection Number	Injection Pressure
4-8	330	133	14 603	6	14.483
4-10	321	121	5 808	3	12.75
5-11	312	118	31 960	12	10.775
6-8	316	138	9 871	4	14.05
6-10	310	126	36 989	14	11.935 7
6-12	304	116	33 810	11	10.872 73
6-14	298	106	40 956	13	10.807 6
7-11	303	125	21 005	8	11.125
7-13	297	117	32 978	12	10.491
8-10	297	134	33 495	12	11.558 33
8-14	287	113	22 888	7	11.857 1
8-18c	271	94	43 137	18	12.361 11
9-11	289	130	27 736	10	11.24
9-13	284	121	24 082	9	11.077 78
9-19	260	92	18 111	7	10.985 7
10-8	294	153	7 262	3	14.366 6
10-10	290	139	35 955	12	11.305
10-14	276	119	43 996	14	10.628 5
10-18	265	101	5 464	2	11.25
10-20	255	87	54 059	18	12.088 89
10-g12	279	134	5 628	3	11.333 33
11-13	273	125	22 909	8	10.775
11-15	266	117	22 185	10	11.13
11-19	253	96	32 833	15	10.906 67

Continued

No.	X-axis	Y-axis	Amount of vapour	Injection Number	Injection Pressure
11–21	245	85	39 198	16	17.275
12–12	271	136	7 201	3	10.7
12–14	262	124	14 842	7	11.3
12–16	257	114	15 139	6	12.666 67
12–18	250	103	48 157	18	11.422 2
12–20	242	90	6 941	2	11.25
12–22	236	81	47 333	18	12.35
13–13	260	131	16 139	8	11.7
13–15	254	122	14 150	6	12.066 67
13–19	242	100	38 296	15	11.386 67
13–21	231	88	36 793	17	11.493 33
13–23	228	80	47 142	18	10.716 67
14–12	259	144	20 009	7	13.3
14–14	252	132	26 886	10	12.87
14–16	247	120	9 499	5	12.76
14–18	238	109	60 406	20	12.02
14–20	230	98	39 853	14	11.785 7
14–22	224	86	45 661	16	12.481 2
14–24	220	76	18 448	10	12.37
15–19	230	105	43 295	16	11.126 67
15–21	227	95	43 681	17	10.6
15–23	215	86	42 702	17	12.070 5
15–25	212	82	37 020	15	11.553 33
15–27	206	65	10 100	5	13.12

Chapter 2　Ordered Weighted Averaging

Continued

No.	X-axis	Y-axis	Amount of vapour	Injection Number	Injection Pressure
16–16	235	128	18 301	7	11.942 8
16–18	227	116	30 489	10	11.12
16–20	221	103	23 346	10	10.91
16–22	215	96	21 915	8	11.575
16–24	205	88	22 070	8	11.757 1
16–26	200	80	13 457	7	12.257 1
16–28	194	62	34 017	11	12
17–21	212	104	41 297	15	10.493 33
17–23	207	94	34 794	15	11.157 1
17–25	198	87	14 418	5	10.66
17–27	190	75	22 961	9	11.711 1
18–16	225	136	26 388	8	12.975
18–18	217	127	67 398	15	12.673 33
18–20	210	112	46 725	16	12.556 25
18–22	203	105	44 697	16	11.606 2
18–24	197	95	44 439	15	12.06
18–26	190	85	58 183	23	12.403
18–28	180	74	26 964	13	12.315 3
18–30	178	62	48 521	20	12.91
19–21	202	112	61 542	14	17.514 2
19–23	192	105	36 373	14	10.35
19–25	187	95	19 379	8	11.775
19–27	180	83	28 534	12	10.608 3
20–18	206	129	32 871	11	13.454 5

Continued

No.	X-axis	Y-axis	Amount of vapour	Injection Number	Injection Pressure
20-20	200	123	57 579	19	11.283 33
20-22	192	112	44 693	16	12.437 5
20-24	186	104	51 608.7	19	12.1
20-26	180	95	50 818	20	11.485
20-28	174	80	14 111	9	11.633 33
20-30	169	73	3 244	1	11.4
20-32	160	58	5 164	2	13.25
21-23	183	107	48 757	19	10.455 56
21-25	183	102	52 788	20	10.018
21-27	173	87	25 643	14	10.621
21-29	164	77	11 808	5	11.84
22-20	188	127	66 217	24	11.165
22-22	177	118	65 106	25	12.496
22-24	175	105	3 286	1	15.2
22-26	170	94	52 529	22	11.309
22-28c	166	85	12 964	6	10.817
22-30	153	71	53 410	22	13.22
22-32	149	64	8 473	4	12.35
22-G24	180	109	49 429	18	10.947
23-25	165	109	29 132	12	11
23-27	161	93	18 304	11	11.236
24-24	166	118	67 009	29	12.025
24-26	162	102	27 695	13	11.675
24-28	151	94	19 661	7	11.5

Continued

No.	X-axis	Y-axis	Amount of vapour	Injection Number	Injection Pressure
24–30	143	78	4 503	2	14.15
24–32	139	70	39 844	17	14.859
24–34	132	60	5 061	2	12.5
24–36	125	50	23 540	10	12.39
25–27	153	105	24 239	11	13.409
26–24	155	120	19 634	9	11.9
26–26	149	112	57 441	20	12.742
26–28	141	99	40 731	15	13.527
26–30	134	88	33 617	14	15.35
26–32	127	77	3 553	2	14.9
26–34	121	65	36 457.9	15	13.913 33
26–36	113	52	5 857	2	14.75
28–28	134	106	53 888	20	13.4
28–30	123	94	19 763	10	16
28–32	119	83	55 383	25	14.528
28–34	112	72	21 033	11	14.72
28–36	104	60	43 429	18	14.43
28–38	98	49	11 471	5	15.733
30–28	120	113	20 126	9	15.644
30–30	113	106	44 696	18	14.762 5
30–32	107	88	29 496	18	14.494
30–34	102	78	48 001	16	15.12
32–26	116	130	9 598	9	17.525
32–32	97	97	19 037	9	14.625

We need to generate raster images from thermal injection data in Table 2-6. The first step is to start IDRISI, and from "Data Entry → Database Workshop" to start database workshop, see Figure 2-12. And the database workshop window is shown as Figure 2-13. Then in database workshop, click New in File menu, and establish a new mdb file as Figure 2-14, Input new mdb file name, such as "vapor number", see Figure 2-15.

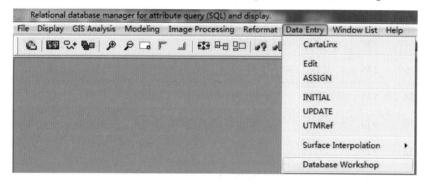

Fig.2-12　Start database Workshop

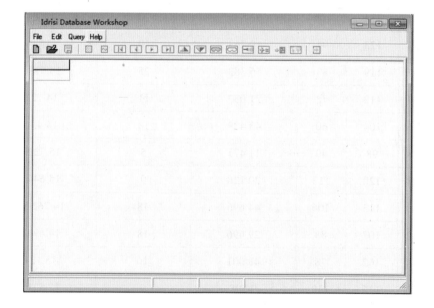

Fig.2-13　Database Workshop window

Chapter 2 Ordered Weighted Averaging

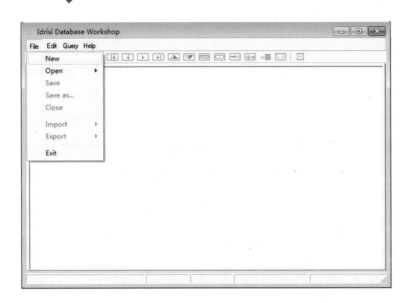

Fig.2-14 Establish a new mdb file

Fig.2-15 Input new mdb file name

In database workshop, from file menu, choose "Import → Table → from External File", click from external file to import, see Figure 2-16.

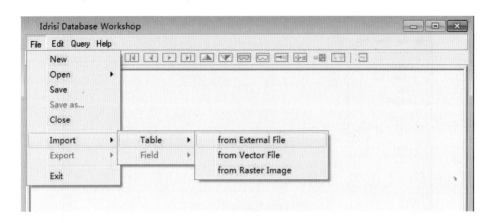

Fig.2-16　Import external file

Import well data to database workshop as Figure 2-17. Click "Open" in Figure 2-17, and pop up windows as Figure 2-18.

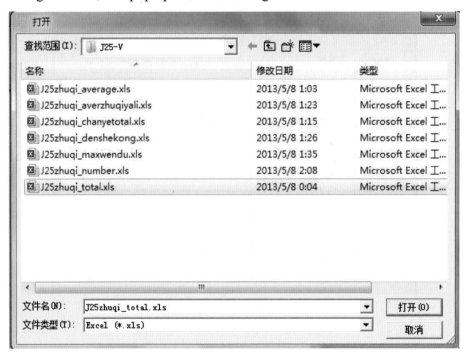

Fig.2-17　Select well data file

Chapter 2　Ordered Weighted Averaging

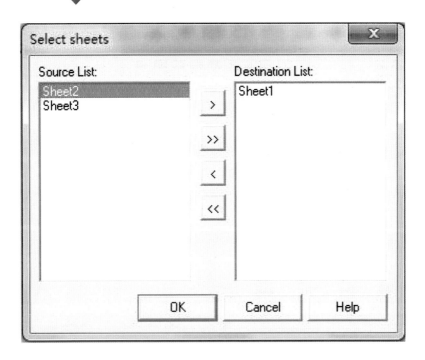

Fig.2-18　Select sheets of well data file

Then click "OK", and click "Yes" in the pop up window that is shown as Figure 2-19.

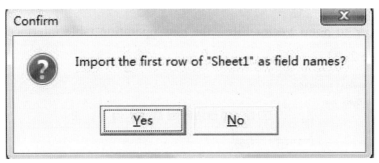

Fig.2-19　Confirm window

Then the amount of vapor injection is import as Figure 2-20.

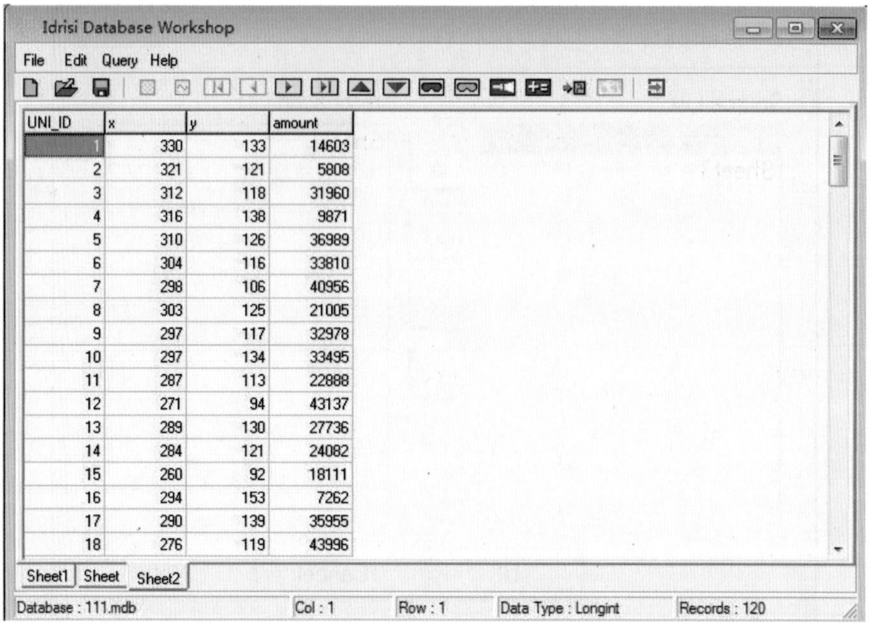

Fig.2-20　Imported well data

After the data input, save the well database file as follows. Click "Save" operation in the file menu can save the data file, as shown in Figure 2-21. Then successively select the "Export" in the file menu and select "X, Y to point Vector File". the new operation table interface shown as Figure 2-22 .

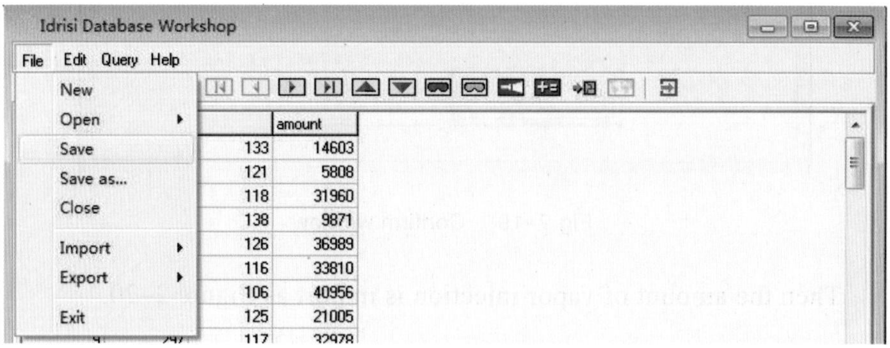

Fig.2-21　Save file

Chapter 2 Ordered Weighted Averaging

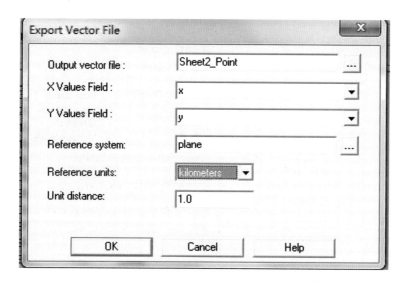

Fig.2-22 Export vector file

Choose x-axis in X Values Field, and y-axis in Y Values Field, and then click OK in export vector file in Figure 2-22. The vector image is shown as Figure 2-23.

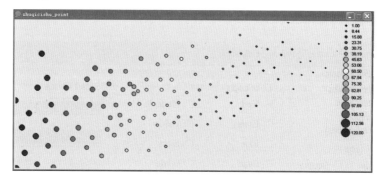

Fig.2-23 Vector file

In the composer interface (Figure 2-24), the layer properties of any layer can be displayed as Figure 2-25. click "Advanced Palette/Symbol Selection" in Figure 2-25. Then in Figure 2-26, select "None (uniform)", and clock OK, the Figure 2-23 can be changed to Figure 2-27.

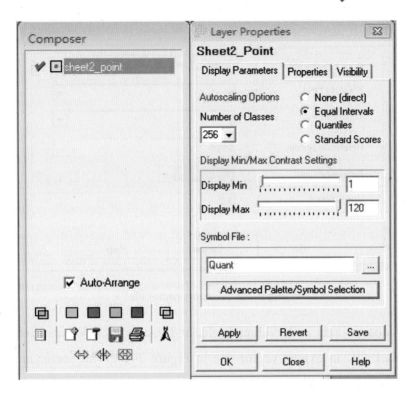

Fig.2-24　Vector file　　Fig.2-25 Vector file

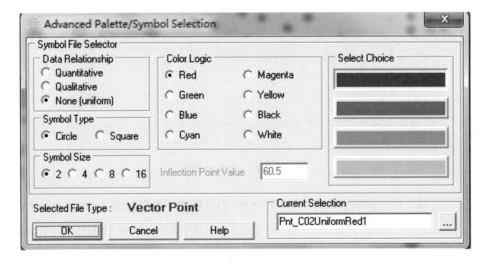

Fig.2-26　Vector file

Chapter 2   Ordered Weighted Averaging

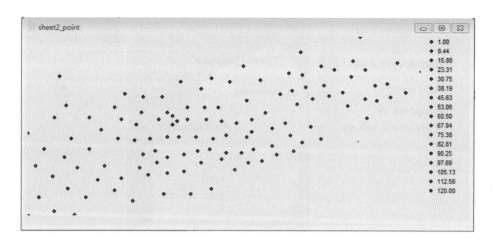

Fig.2-27   Vector (x,y) file

Then establish a display link, click the button of "establish a display link" (reference to Figure 2-20 or 2-21), and shown as Figure 2-28. Please pay attention to Figure 2-28, the Link field name must choose "UNI_ID". In database workshop, choose File → Export → Field → to vector file, see Figure 2-29.

Fig.2-27   Establish a display link

基于GIS的决策支持与表面分析
GIS-Based Decision Support & Surface Analysis

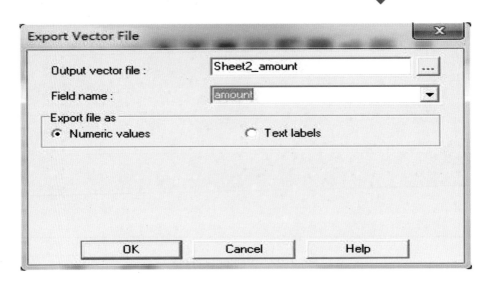

Fig.2-29　Export vector file

Until now, we obtain the vector image of "amount of vapor", shown as Figure 2-30.

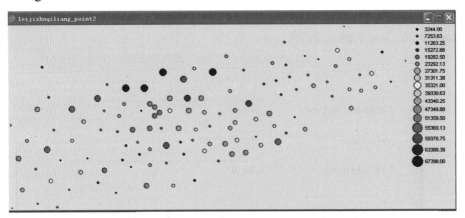

Fig.2-30　Vector image of vapor amount

In the same way, vector images of injection pressure and injection number are obtained as Figure 2-31 and Figure 2-32.

Chapter 2 Ordered Weighted Averaging

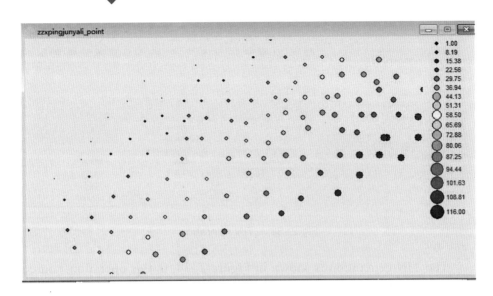

Fig.2-31　Vector image of injection pressure

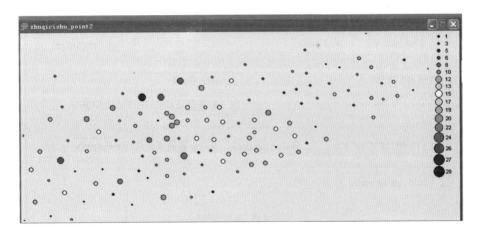

Fig.2-32　Vector image of injection number

Now, we need to generate raster images through TIN operation, see Figure 2-33.

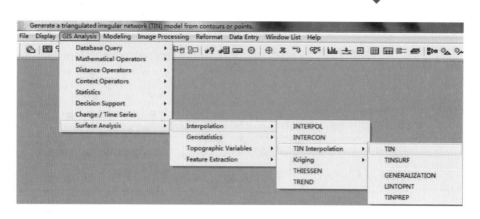

Fig.2-33　TIN operation

The TIN generation is shown in Figure 2-34, in this operation, source is points, and select input vector file and out TIN file.

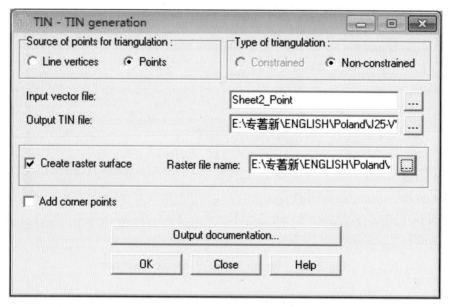

Fig.2-34　TIN generation

When click "OK" in Figure 2-34, the pop up TINSURF window is shown as Figure 2-35.

Chapter 2   Ordered Weighted Averaging

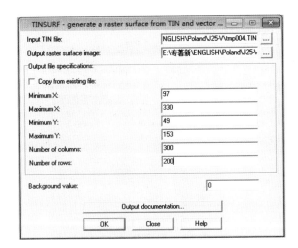

Fig.2-35   TINSURF

In Figure 2-35, the number of columns and rows should be input, for example 300 and 200 in this instance, and the raster image is generated as Figure 2-36.

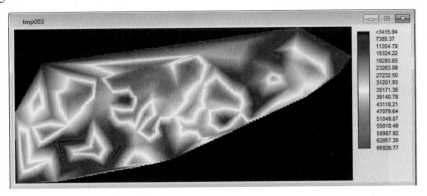

Fig.2-36   Raster image of vapor amounts

The FUZZY module is shown as Figure 2-37. In this instance, linear function type can be selected, and output data format is selected as Byte [0-255 range]. Especially, control point "a" and "b" must be input as MIN and MAX value of vapor injection amounts. FUZZY images of three factors are shown from Figure 2-38 to Figure 2-40.

Fig.2-37　FUZZY operation

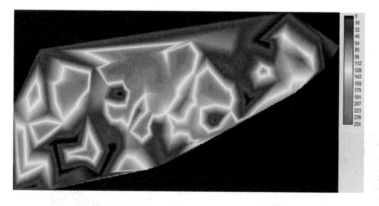

Fig.2-38　Suitability of vapor amounts

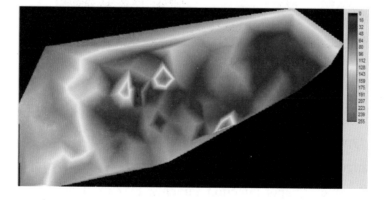

Fig.2-39　Suitability of injection pressure

Chapter 2 Ordered Weighted Averaging

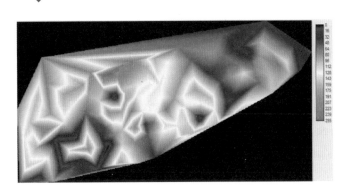

Fig.2-40 Suitability of injection number

Next, before evaluation, criteria weights must be calculated firstly. In WEIGHT module, corresponding comparison matrix can be constructed. Each value of comparison matrix reflects the relative importance of factors. Comparison matrix is shown as Table 2-7.

Table 2-7 Comparison matrix

Factor	A_1	A_2	A_3
A_1	1	1/3	1/5
A_2	3	1	1/2
A_3	5	2	1

If comparison matrix is constructed in WEIGHT module, the number of factors and the file name of standardized images must be specified. The comparison matrix produced by this technique, is a positive reciprocal matrix. Therefore, only the higher/lower triangular half which includes $n(n-1)/2$ elements needs to be filled in.

The results of criterion weights for vapor injection amounts, injection pressure, injection number are 0.1095, 0.3090, and 0.5816. Consistency Ratio is 0.00, it is less than 0.10, and indicate good consistency. If Formula (2-8) is used to calculate order weights, the results are 0.167, 0.333, and 0.5. Finally, choose MCE module in decision support of GIS analysis menu, and pick Order weighted average in MCE window, see Figure 2–41.

Choose number of factors as 3, number of constraints as 0. Then make choice of factor filename and factor weights for those three factors, especially, three order weights should be input, of course, output image and title also should be inputted. Click "OK" for OWA evaluation. The evaluation result is shown in Figure 2–42.

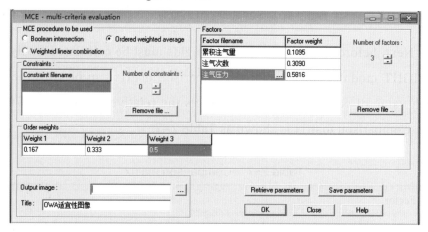

Fig.2–41 OWA operation in MCE

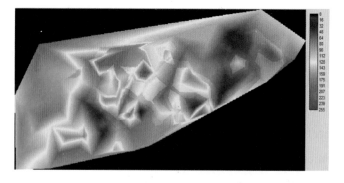

Fig.2–42 Risk of casing failure

From Figure 2–42, we can find high risk areas (dangerous areas) in J25 block, and more attention should be paid to those dangerous areas.

Chapter 3   Multiple Objectives

Sometimes, it is necessary to make decisions that satisfy several objectives, this is called multiple objectives decision making.

GIS has become a necessary system in decision-making, and multiple objective decisions making is common in environmental management, but it is not yet further developed within GIS. With conflicting objectives, land can be allocated to one objective but not more than one. One possible solution lies with a prioritization of objectives. After the objectives have been ordered according to priority, the needs of higher priority objectives are satisfied before those of lower priority ones. This is done by successively satisfying the needs of higher priority objectives and then removing areas taken by that objective from consideration by all remaining objectives. More often a compromise solution is required. Compromise solutions to the multi-objective problem have most commonly been approached through the use of mathematical programming tools outside GIS. In the case of raster GIS, a solution to the problem of multi-objective problems under conditions of conflicting objectives is very important to discuss.

3.1 Conflict Objectives Analysis

The procedure for multi-objective problems is an extension of single objective problems; it is illustrated by the diagram in Figure 3-1.

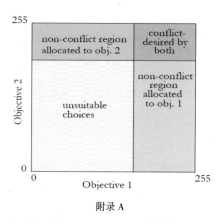

附录 A

Fig. 3-1 Procedure for multi-objective

In Figure 3-1, each of the suitability images may be regarded as an axis in a multi-dimensional space. Here two objectives of simple explanation are considered. Every raster cell in the image can be located within this decision space according to its suitability level on each of the objectives. To find the best area for Objective 1, it is simply need to move a decision line down from the top (i.e., far right) of the Objective 1 suitability axis until enough of the best raster cells are captured to meet our area target. The same with the Objective 2 can be done. As can be seen in Figure 3-1, this partitions the decision space into four regions — areas best for Objective 1 and not suitable for Objective 2, areas best for Objective 2 and not suitable for Objective 1, areas not suitable for either, and areas judged best for both. The latter represents areas of conflict.

To resolve these areas of conflict, the decision space can also be partitioned into two further regions: those closer to the ideal point for Objective 1 and those closer to that for Objective 2. The ideal point represents the best possible case — a cell that is maximally suited for one objective and minimally suited for anything else. Since the conflict region will be divided between the objectives, both objectives will be short on achieving

Chapter 3 Multiple Objectives

their area goals. As a result, the process will be repeated for both objectives to gain more territory. The process of resolving conflicts is iteratively repeated until the exact area targets are achieved. However, unequal weighting can be given, which has the effect of changing the ratio of the weights assigned to those objectives. It should also be noted that just as it is necessary to standardize criteria for multi-criteria evaluation, it is also required for multi-objective evaluation.

The conflict area can be more divided into two areas, one for Objective 1, and another for Objective 2. Different weights among objectives will change the borders, and more suitability area will divide to objective with more weight. This procedure is easily fulfilled with MOLA module in IDRISI, see Figure 3-2. And the pop up MOLA window is shown as Figure 3-3.

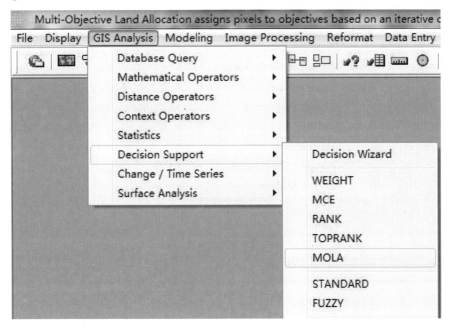

Fig. 3-2 MOLA in IDRISI

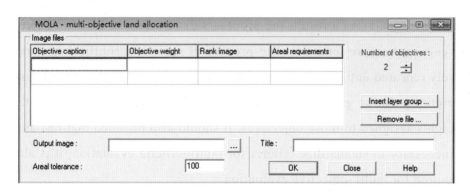

Fig. 3-3 MOLA window

From Figure 3-3, we can notice that Objective weight and Areal requirements for each objective should be input, and images of all objectives must be Rank image. We can obtain rank images from rank module, see Figure 3-4, and Figure 3-5.

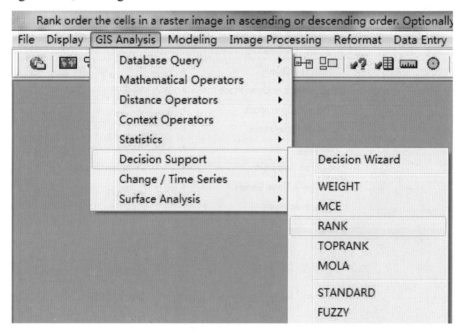

Fig. 3-4 RANK in IDRISI

Chapter 3 Multiple Objectives

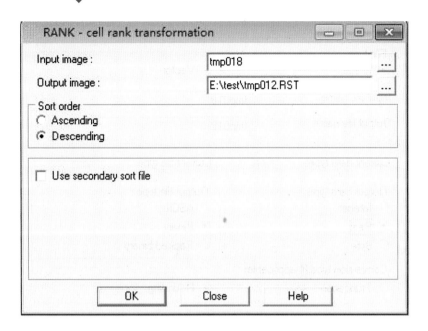

Fig. 3-5 RANK window

In Figure 3-5, the cell rank transformation needs Binary type image as input image, this can be fulfilled by CONVERT module in IDRISI, see Figure 3-6. The pop up CONVERT window is shown as Figure 3-7.

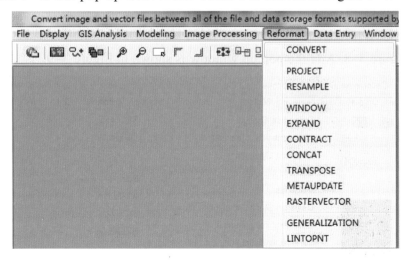

Fig. 3-6 CONVERT in IDRISI

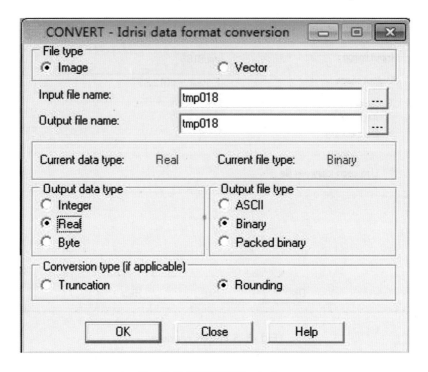

Fig. 3-7 CONVERT window

The orange input file is suitability image come from MCE evaluation as single objective evaluation.

The conflict objectives can be displaced with CROSSTAB module, see Figure 3-8. And the pop up window is shown in Figure 3-9.

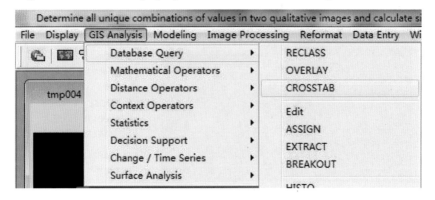

Fig. 3-8 CROSSTAB in IDRISI

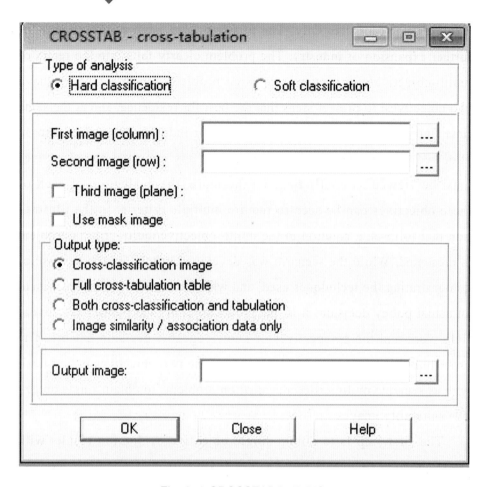

Fig. 3-9 CROSSTAB in IDRISI

Therefore, for conflict objectives problem, multiple objectives evaluation can be fulfilled through MOLA module, displaced and analyzed through CROSSTAB module. Next, we will introduce application example of multiple objectives.

3.2 Application Examples

As a multi-criteria/multi-objective procedure, we consider the following example of developing a zoning map to industry within construction areas in Tangshan City. The problem is to zone suitable areas for further expansion of industry. The problem clearly falls into the realm of multi-objective/multi-criteria decision problems. In this case, we have two objectives: to protect areas that are best for residence, and at the same time find other lands that are best suited for industry. Since land can be allocated to only one of these objectives at any one time, the objectives must be viewed as conflicting. Furthermore, the evaluation of each of these objectives can be seen to require multiple criteria. In the illustration that follows, a solution to the multi-objective/multi-criteria problem is presented. While the scenario was developed purely for the purpose of demonstrating the techniques used, and while the result does not represent an actual policy decision, it is one procedure that incorporates substantial field work and the perspectives of knowledgeable decision makers. The procedure follows a logic in which each of the two objectives is first dealt with as separate multi-criteria evaluation problems to obtain single objective suitability images

The first step is to obtain single objective evaluation results with the methods in Chapter 2 and Chapter 1. Here, it is assumed that you are familiar with the previous methods, and we have worked out the single objective suitability images for suitable residences and industry areas. Of course, the single objective multi-criteria evaluations are solved through establishing the criteria, standardizing the factors, establishing factor weights and undertaking the multi-criteria evaluation. Then, those two maps are compared to arrive at a single solution that balances the needs of the two competing objectives. Those two images are shown as Figure 3-10 and Figure 3-11. The constraint condition is underground pipeline.

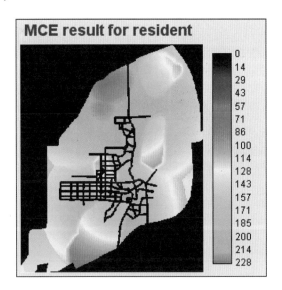

Fig.3-10　Suitability image for resident

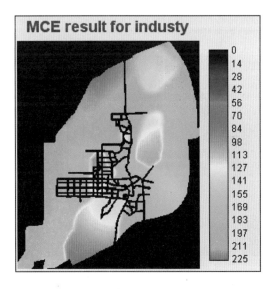

Fig.3-11　Suitability image for industry

The next step is to standard those images to rank images (reference to Figure 3-4 and Figure 3-5). Rank images that obtained from rank module are shown as Figure 3-12 to Figure 3-14.

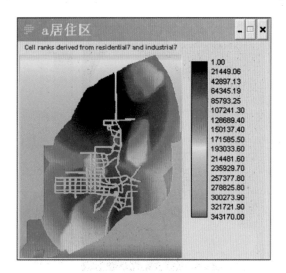

Fig.3-12　Rank image for residence

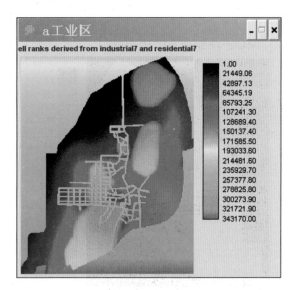

Fig.3-13　Rank image for industry

Chapter 3 Multiple Objectives

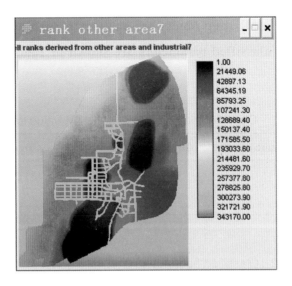

Fig.3-14 Rank image of restricted development

Once the multi-criteria suitability maps have been created for each objective, the multi-objective decision problem can be approached. The suitable areas can be assigned through RECLASS or decision wizard, those are shown as follows.

Each objective image for this analysis can be obtained from RECLASS operation, see Figure 3-15.

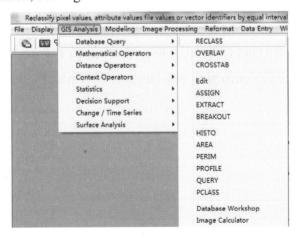

Fig.3-15 RECLASS module

Images for conflicting objectives analysis is fulfilled in RECLASS module and the classified parameters are shown in Table 3-1 and Table 3-2.

Table 3-1   Classified parameters of suitability value for resident

Assign a new value of	To all values from	To just less than
0	100 000	5 000 000
1	1	100 000

Table 3-2   Classified parameters of suitability value for industry

Assign a new value of	To all values from	To just less than
0	1	20 000
1	20 000	150 000
0	150 000	500 000

The results are shown as Figure 3-16 and Figure 3-17. Therefore, conflicting objectives can be analyzed from RECLASS operation with Figure 3-16 and Figure 3-17.

Then, we analyze the objective conflicting. The conflict objectives are analyzed with CROSSTAB module (reference to Figure 3-8 and Figure 3-9). The image of conflicting objectives is shown as Figure 3-18.

Chapter 3　Multiple Objectives

Fig. 3-16 Resident areas

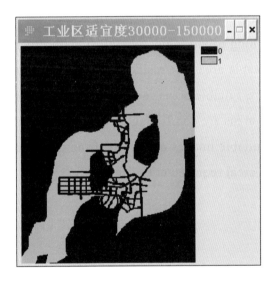

Fig. 3-17 Suitability industry areas

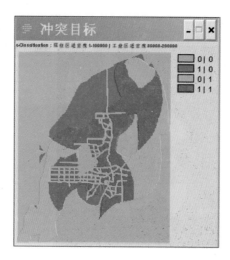

Fig.3-18　Image of conflicting objectives

In Figure 3-18, there are four types of areas. The first areas are areas that do not belong to any objective (areas with green color). The second areas are suitable areas for residence that are displayed in blue color. The third areas are suitable areas for industry that are displayed in yellow color. The fourth areas are conflicting areas that are displayed in red color.

Of course, suitable images may come from decision wizard operation through assigned areal requirements, see Figure 3-19.

Chapter 3 Multiple Objectives

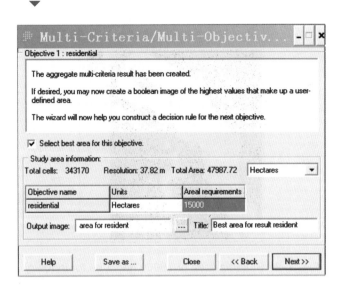

Fig. 3-19 Areal requirements for resident

And the results are shown as Figure 3-20 and Figure 3-21.

Fig. 3-20 Suitable area of 15 000 hectares for resident

Fig. 3-21 Suitable area of 15000 hectares for industry

Here, we can guess the difference between RECLASS and decision wizard for this operation.

Now, we use MOLA module to finish our evaluation. MOLA requires the names of the objectives, the relative weight to assign to each, and the area to be allocated to each (reference to Figure 3-3). The module then undertakes the iterative procedure of allocating the best ranked cells to each objective according to the areal goals, and resolving conflicts based on the weighed minimum-distance-to-ideal- point logic. Different results come from different weights are investigated as follows.

For instance, as Figure 3-22, equal weight is assigned to residence and industry area with 0.5, and 0 for restricted construction areas. Areal requirements for residence and industry areas are both 70 000 hectares.

Chapter 3 Multiple Objectives

Fig. 3-22 Weights decision analysis of MOLA (0.5)

The result is shown in Figure 3-22. It can be found in Figure 3-22 that restricted construction areas are not affected by weight if the areal requirements for residence and industry are not very much. Since restricted construction areas are low suitability areas.

Fig.3-23　Result of MOLA (0.5)

In Figure 3-24, weight for residence is assigned 1.0, and both 0 for industry and restricted construction areas. Areal requirements for residence and industry areas are both 70 000 hectares, and 0 for restricted construction areas. The results are shown in Figure 3-25.

Fig. 3-24 Weights decision analysis (1.0)

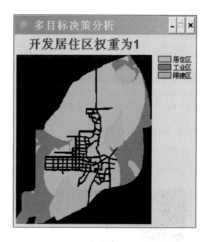

Fig.3-25 Weights decision analysis of MLOA (1.0)

It can be found in Figure 3-22 that restrict construction and industry areas are both reserved although their weights are 0, even areal requirements for restrict construction are also 0. It means that when areas are enough to all objectives, all objectives can be satisfied even though NOT assign weights and areal requirements for them.

In Figure 3-26, weight for residence is assigned 0.7, 0.2 for industry and 0.1 for restrict construction areas. Areal requirements for residence are 100 000 hectares, 50 000 hectares for industry, and 30 000 hectares for restrict construction areas. The results are shown in Figure 3-27.

Chapter 3　Multiple Objectives

Fig.3-26　Weights decision analysis (0.7)

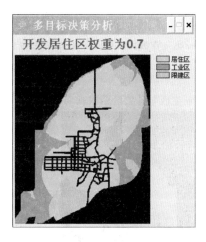

Fig.3-27　Weights decision analysis of MLOA (0.7)

From Figure 3-27, it can be found that more areas are assigned to residence, and both restrict construction and industry areas are reserved. This shows that weights for different objectives will change the distribution (location assign), but all objectives may be satisfied if there are enough areas to assign for them. If we compare of those three images (Figure 3-23, Figure 3-25, and Figure 3-27), we can found that the best suitable residence areas are located in the north-west of Tangshan City, although distribution man be change.

In Figure 3-28, weight for residence is assigned 0.2, 0.1 for industry and 0.7 for restrict construction areas. Areal requirements for residence, industry and restrict construction areas are both 60 000 hectares. The results are shown in Figure 3-29.

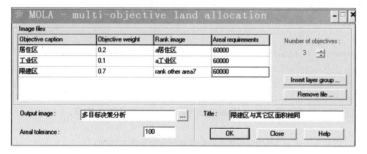

Fig.3-28   Weights decision analysis (0.2)

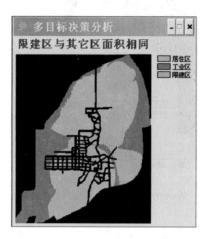

Fig.3-29   Weights decision analysis of MLOA

In Figure 3-29, there is maximum area for restrict construction. It comes from the weights and areal requirements are assigned.

From above analysis, it can be found that the most important for multiple objectives is there are enough areas for all objectives. Weights affect the distribution of each objective, and the most safety way is to assign constraint factor an enough weight and proper areal requirements. In fact, we think the most important factor is constraint, and then residence factor.

Chapter 3 Multiple Objectives

Therefore, we consider constraint factor the first need to be satisfied factor, then residence, and last for industry.

Through AREA module, we can check the results of multiple objectives; it is shown as Figure 3-30.

Fig.3-30 Area calculation of MLOA

From Figure 3-30, we can check the results of category, size of assigned area for each objective. In this instance, there are more than 20000 hectares residue areas. Therefore, all objectives are easy to be satisfied. Here, we can guess the different between single and multiple objectives decision.

For single objective decision, the original image is suitability image, and for multiple objectives, are ranking images. Therefore, less risk is considered in multiple objectives decision making, thus, the constraint factor should be the first factor, and areal requirements must be assigned properly. In other word, once the multi-criteria suitability maps have been created for each objective, the multi-objective decision problem can be approached. The conflicts map is shown as Figure 3-18. Single objective decision making is the basis of multiple objectives decision making. In multiple objectives problem, the first step is to rank order the cells in each of the suitability maps. This prepares the data for use with the MOLA procedure and has the additional effect of standardizing the suitability maps using a nonparametric histogram equalization technique. This preserves the basic logic of the uncorrelated ideal points for conflicting objectives that is used in the resolution of conflicts. The second step was to submit the ranked suitability maps to the MOLA procedure. MOLA requires the names of the objectives, the relative weight to assign to each, and the area to be allocated to each. The module then undertakes the iterative procedure of allocating the best ranked cells to each objective according to the areal goals, and resolving conflicts based on the weighed minimum-distance-to-ideal-point logic. Ordinal, OWA method is applied to the suitability evaluation single objective. This process can also apply to other fields of decision analysis.

We can summarized multi-objective decision problem as follows: the single objective multi-criteria evaluations are solved through establishing the criteria, standardizing the factors, establishing factor weights and undertaking the multi-criteria evaluation. The multi-objective problem is solved through standardizing the single-objective suitability maps and allocating the best ranked cells to each objective.

Finally, some advice is proposed to land use planning in Tangshan City based on multiple objectives evaluations. Suitability increases from southeast to northwest, and the most favorite areas located in the northwest of Tangshan City. Some old residential areas locate in restrict construction areas, and more attentions must be paid on those areas.

Prat II Surface Analysis

Chapter 4　Uncertainty Analysis

Uncertainty includes any errors, unclear or variation in database or decision rules. In the procedure of decision making, uncertainty can occur in the evidence, the decision set, and the relation that associates them. In MCE evaluation, uncertainty comes from at least three sources. The first source is the definition of a criterion results in uncertainty. The second source is the degree of the factor data to support the decision. The third source is the relation between the evidence and the decision set.

In the second source, similar to fuzzy set membership functions, belief are used to represents the degree of evidence suggests decision. Thus Dempster–Shafer theory is used in this chapter for this analysis. The Dempster–Shafer theory provides an important approach for indirect evidence and incomplete information, and it is another method to combine data. In IDRISI, the Belief module is used to combine much different information to predict the probability. In other word, it represents the relative risk of decision making according to the degree of available of the total information.

4.1 Dempster-Shafer Theory

Dempster–Shafer theory is an extension of Bayesian probability theory, and has been widely applied in many fields. The basic of Dempster–Shafer theory is to establish frame of discernment Ω, which is the set of all possible results of all incompatible proposition. If there are three basic hypotheses {A, B, C} in a frame of discernment, then all combinations of A, B, and C, are evidences in Dempster-Shafer theory, which include [A], [B], [C], [A,B], [A,C], [B,C], and [A,B,C].

The first three singleton hypotheses include only one basic element. Non-singleton hypotheses include more than one basic element. These hierarchical combinations are very important since it is common that some combinations of hypotheses are supported by the evidence but not further subsets. For example, there are two classes of land cover classification: deciduous and conifer, it is easy to distinguish forest from non-forested areas, but difficult to distinguish deciduous and conifer. In this case, the hierarchical combination [deciduous, coniferous] can be used as combinations supported by evidence. Although it is a statement of uncertainty, it is valuable to Dempster-Shafer to statement of belief to the hierarchical combination.

In Dempster-Shafer theory, a basic probability assignment (BPA) represents the support that a piece of evidence provides for one of these hypotheses and not its proper subsets. This is usually symbolized with the letter "m" to $m(A)$. That is m: $2^\Omega \rightarrow [0, 1]$, with,

$$\begin{cases} m(\phi) = 0 \\ \sum_{A \subset 2^\Omega} m(A) = 1 \end{cases} \quad (4-1)$$

In which,

ϕ = empty set,

$m(A)$= basic probability assignment A.

Belief represents the total support for a hypothesis, and belief function is defined as,

$$\text{Bel}(X) = \sum_{Y \subseteq A} m(Y) (\forall X \subset \Omega, Y \subset 2^\Omega) \quad (4-2)$$

Chapter 4 Uncertainty Analysis

In which,

Bel(X) =belief function,

Bel(ϕ) =0, and Bel(Ω) =1.

If m (A) >0, then A is the core of belief function. Bel($\overline{A}$) represents the degree of NOT A.

Thus the belief in [A, B] will be calculated as the sum of the BPAs for [A, B], [A], and [B].

In contrast to belief, plausibility represents the degree to which a hypothesis cannot be disbelieved. Plausibility function is defined as,

$$\text{Pl}(X) = 1 - \text{Bel}(\overline{X}) = \sum_{X \cap Y \neq \phi} m(Y)(\forall A \subset \Omega, Y \subset 2^{\Omega}) \quad (4-3)$$

In which, Pl(X) is the degree to which the conditions appear to be right for that hypothesis, even though hard evidence is lacking. And there is Bel(X)≤Pl(X). For each hypothesis, belief is the lower boundary of our commitment to that hypothesis, and plausibility represents the upper boundary. This is shown as,

$$\text{Bel}(X) \leqslant P(X) \leqslant \text{Pl}(X) \quad (4-4)$$

Dempster-Shafer theory is an extension of ordinary probability theory, Pl(X) − Bel(X) represents uncertainty, which is shown as Figure 4-1.

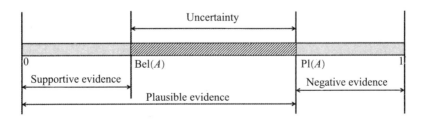

Fig.4-1 Uncertainty intervals of evidence theory

The Dempster-Shafer rule of combination provides an important approach to aggregating indirect evidence and incomplete information. If m_1

and m_2 is two incompatible basic probability assignments, a new m function based on $m(A) = m_1 \oplus m_2$ is expressed as follow,

$$\begin{cases} m(A) = m_1 \oplus m_2 = \dfrac{\sum\limits_{B_i \cap C_j = A} m_1(B_i) m_2(C_j) / K}{K = 1 - \sum\limits_{B_i \cap C_j = \phi} m_1(B_i) m_2(C_j)} \end{cases} (A \neq \phi) \qquad (4-5)$$

In which,

B_i = the i-th focus element of belief function Bel_1,

C_j = the j-th focus element of belief function Bel_2.

If $K \neq 0$, then $m(A)$ is the sum of m_1 and m_2, otherwise, if $K=0$, $m(A)$ is not exist, it means confliction of m_1 and m_2.

If there are n basic probability assignment, a new combinative basic probability assignments $m(A) = m_1 \oplus m_2 \oplus \ldots \oplus m_n$ is expresses as,

$$\begin{cases} m(A) = m_1 \oplus m_2 \oplus \ldots \oplus m_n = \dfrac{\sum\limits_{\cap A_{ij} = A} \bigcap\limits_{1 \leq i \leq n} m_i(A_{ij}) / K}{K = 1 - \sum\limits_{\cap A_{ij} = \phi} \bigcap\limits_{1 \leq i \leq n} m_i(A_{ij})} \end{cases} (A \neq \phi) \qquad (4-6)$$

In which,

A_{ij} = the j-th focus element of belief function Bel_i,

If $K \neq 0$, then $m(A)$ is the sum of $m_1, m_2, \ldots, m_n$, if $K=0$, them it is conflictive among $m_1, m_2, \ldots, m_n$, and $m(A)$ is not exist.

In contrast to belief, plausibility represents the degree to which an hypothesis cannot be disbelieved. Unlike the case in Bayesian probability theory, disbelief is not automatically the complement of belief, but rather, represents the degree of support for all hypotheses that do not intersect with that hypothesis. Thus,

$$Pl(X) = 1 - Bel(\bar{X}) \quad \text{where} \quad \bar{x} = not\ X \qquad (4-7)$$

Thus, we can obtain the follow,

$$Pl(X) = \sum m(Y) \quad \text{when} \quad Y \cap X \neq \phi \qquad (4-8)$$

Chapter 4 Uncertainty Analysis

In summary, belief is the degree of evidence to support a hypothesis; plausibility represents the degree to which the evidence does not refute that hypothesis. Therefore, belief and plausibility can be treated as the lower and upper boundary of uncertainty, belief interval indicate the degree of uncertainty or a measure of uncertainty about a specific hypothesis, which is the extent between the lower and upper boundary. New evidences should be supplied to high belief interval areas to improve the degree of evidence supporting.

Dempster-Shafer theory is a very useful tool to reduce uncertainty through discernment frame establishment and data combination. In IDRISI, Dempster-Shafer procedure can be fulfilled by Belief module, see Figure 4-2.

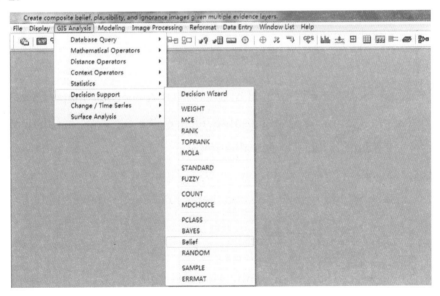

Fig.4-2 Belief operation in IDRISI

Frame of discernment of Dempster-Shafer theory can be built in Belief module for the full hierarchy. With Belief module, images of the belief, plausibility or belief interval associated for a hypothesis can be worked out through combination evidences; Belief can be treated as a very important tool to combine direct and indirect evidence with the Dempster-Shafer theory. Belief module window is shown as Figure 4-3.

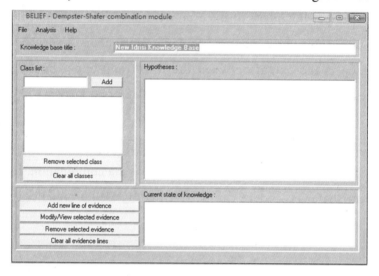

Fig.4-3 Belief window

Dempster-Shafer is method of MCE/MOLA decision support to combine data that known as weight-of-evidence. We can consider the application of Belief in land use evaluation; the decision frame includes two basic elements, [safety] and [non-safety]. Three images of belief, plausibility and belief interval will be obtained by the application of Belief.

4.2 An Application Example

Now, we introduce an application example of Belief module. The Advantage of Belief module is to calculate the probability with different sources information. Belief module provides a tool to establish belief values with incomplete information. Therefore, it makes the decision risk analysis based on information completeness and certainty becomes possible. Expert knowledge is used to transform the evidence into risk or uncertainty probabilities, and the results are probability surfaces that are factors to be combined in Belief module.

In Belief module, all possible hypotheses and their hierarchical structures are including in a frame of discernment or a frame of decisions. With the frame of decisions and basic probability assignments, belief, plausibility, and belief interval images for each hypothesis can be created. The functions of Belief module include: create files of the hierarchical combinations of hypotheses, build basic probability assignments for all hypotheses based on evidences, and display summary images. It is an evaluation of what is known and what is not known.

Belief module requires creating an entire knowledge of all hypotheses firstly than to build summary information. It needs to develop basic probability assignment images with real number values between 0 and 1, which indicate the degree of evidence to support a hypothesis.

For example, we evaluate the safety of site in Tangshan city, there are two hypotheses: [safety] and [non safety], it is the simplest case. Of course, more than two hypotheses are also possible in which there are a group of classes.

We start Belief module firstly (reference Figure 4–3), enter a name for a new knowledge base, such as "site evaluation in Tangshan City", see Figure 4–4. Then, in the Class List textbox, enter each basic class and click on the Add button, all possible classes here are safety and non safety. And three hypotheses will displace in the hypotheses area, they are [safety], [non safety], and [safety, non safety]. You can remove any class by clicking the Remove selected class button. Remove all classes by clicking the Clear all classes button.

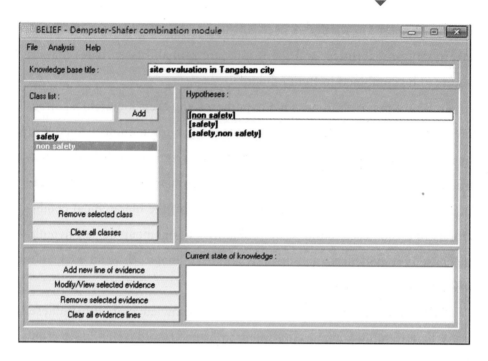

Fig.4-4　enter each basic class in Belief

In this instance, the basic classes include safety and non safety. When safety is entered in the class list, [safety] will be built in the Hypotheses list. When non safety is entered as a second input, the Hypotheses list will show all hypotheses: [no safety] and [safety, non safety]. After all hypotheses are inputted, the next step is to add evidences. Click the Add new line of evidence button to launch the Evidence Editor, it is shown as Figure 4-5.

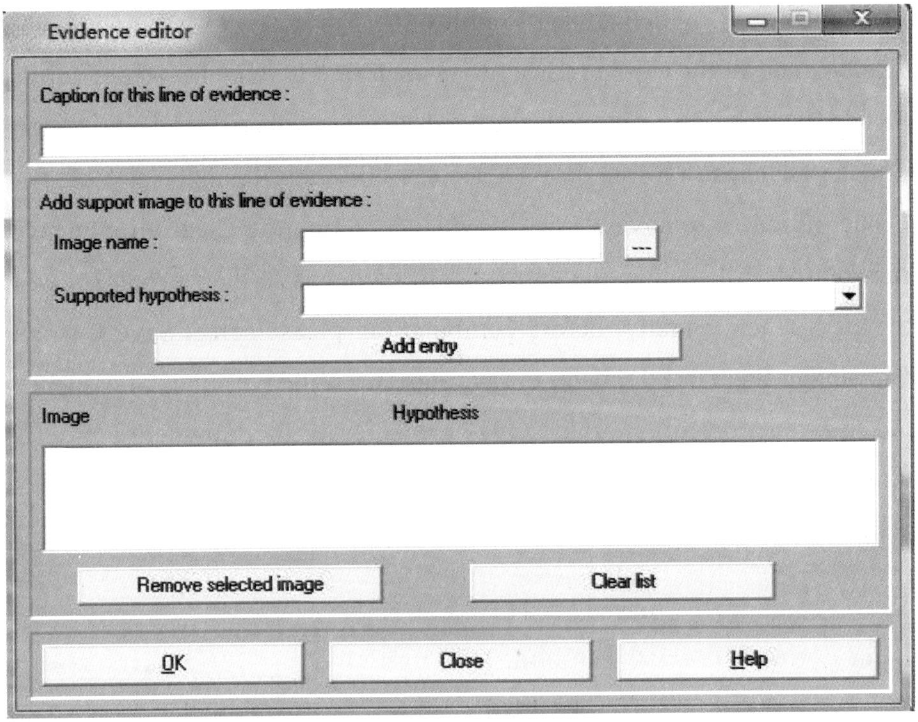

Fig.4-5 Evidence editor window

Enter a caption for the new line of evidence, such as earthquake in this case. This is descriptive information that will display in the Current state of knowledge box in the Belief dialog. Select image name and choose the hypothesis that this image supports in Supported hypothesis box. Click the Add entry button to add the image file and supported hypothesis. Notice that you can input all evidences to support this hypothesis, and all images must have real number value between 0 and 1. Repeat this procedure, and add all additional files and supported hypotheses in the same fashion. Of course, you can remove or clear list through click the Remove selected image button or the Clear list button.

Click OK to return to the main Belief dialog, and repeat this process for each line of evidence, see Figure 4-6. The current state of knowledge box shows the captions describing each line of evidence added to the knowledge base. You can click the Modify/View selected evidence button to make changes to a line of evidence. Or click the Remove selected evidence button to remove a line of evidence. Click the Clear all evidence lines button will remove the entire list of evidence in the Current state of knowledge box. Finally, before combination, please select Save Current Knowledge Base in File menu to save this file. When all lines of evidence and their supported hypothesis files are entered, the knowledge base is complete, it is ready for combination.

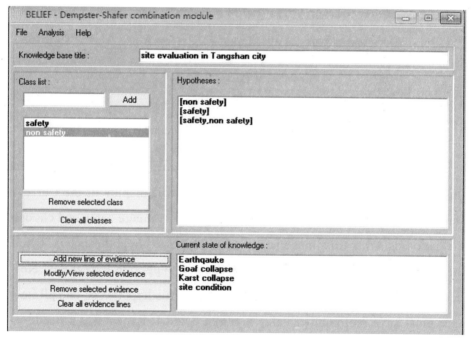

Fig.4-6 The main Belief dialog

Choose Build Knowledge Base Enter in the Analysis menu to calculate the result, see Figure 4-7.

Chapter 4 Uncertainty Analysis

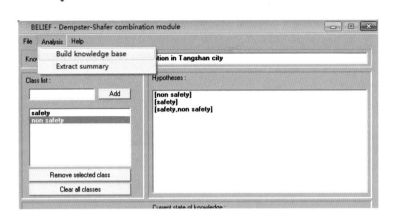

Fig.4-7   Build knowledge base

Enter the Analysis menu and select Extract summary to launch a new dialog, see Figure 4-8. Choose a hypothesis and check the option of Belief, Plausibility, and Belief Interval images for the selected hypothesis. Enter output filenames and click OK. Images are shown as Figure 4-9 to Figure 4-11. Extract summary only extracts files for one hypothesis at a time. To extract images for other hypotheses, simply select Extract summary again, and highlight the desired hypothesis to summarize.

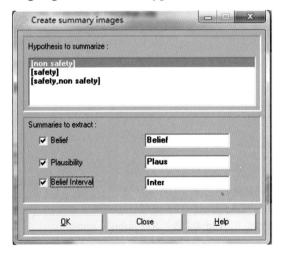

Fig.4-8   Create summary images

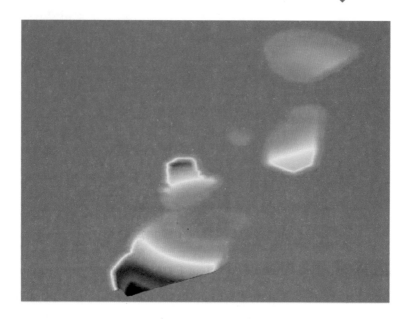

Fig.4-9　Belief image

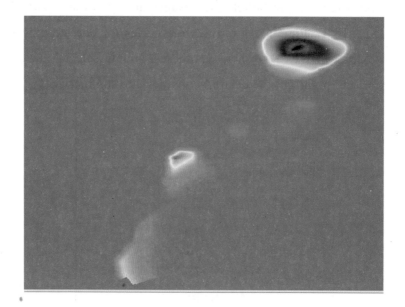

Fig.4-10　Plausibility image

Chapter 4　Uncertainty Analysis

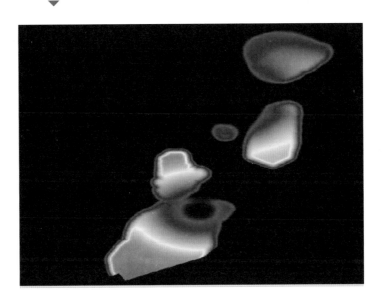

Fig. 4-11 Belief interval image

In order to check all those images together, collection editor is used to group those three images, see Figure 4-12, and Figure 4-13.

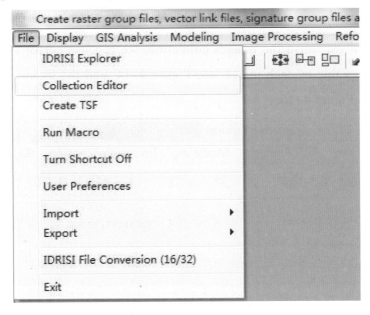

Fig. 4-12 Start collection editor

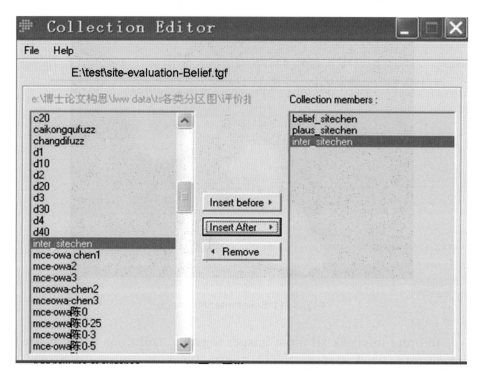

Fig. 4-13 Collection editor window

When we finish the images collection, we can use cursor inquiry to check the value of each image, see Figure 4-14, and the inquiry result is shown as Figure 4-15.

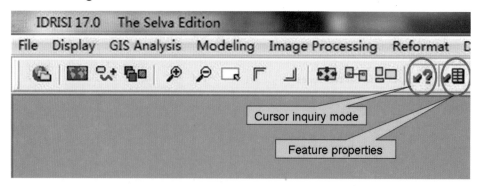

Fig. 4-14 Collection editor window

Chapter 4 Uncertainty Analysis

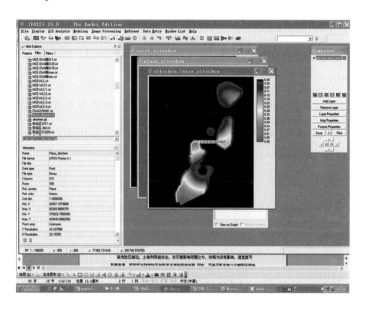

Fig. 4-15 Feature Query

With feature properties mode, the attributes of those three images at any point can be displayed as Figure 4-16.

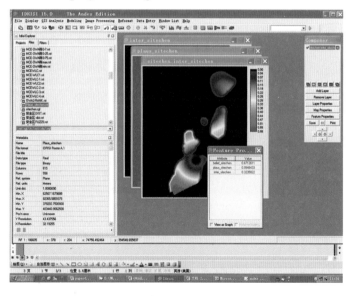

Fig. 4-16 Relation among belief, plausibility, and belief interval

111

The plausibility image (reference to Figure 4-10) shows the degree to which the evidence does NOT refute the hypothesis. Even belief values are low; it is still possible to have high plausibility values. The belief interval is the difference between belief and plausibility, and it is a measure of uncertainty. With high value of belief interval, high plausibility values may indicate that more evidence should be collected.

In this application example, the Belief module is used to fulfill the Dempster-Shafer logic. Three images of belief, plausibility and belief interval are created. The decision frame includes two basic elements, [safety] and [non-safety]. The relation of Plausibility-belief=belief interval is proved. And it also shows that Dempster-Shafer theory is a very useful tool to combine indirect evidence and incomplete information. Also, there are many uncertainty areas with high plausibility and high belief interval, more attention should be paid to those areas.

Chapter 5   Geostatistics

Geostatistics is an important tool to analyze the spatial data values and their locations. Spatial variability is the focus of geostatistics in GIS analysis. Spatial variability present in the sample data is assessed in terms of distance and direction, and can be described as surfaces. There are three modules in IDRISI: Spatial Dependence Modeler, Model Fitting, and Kriging and Simulation. Many tools are provided in geostatistics for spatial dependency analysis of sample data, for example, neighboring points have similar attributes. Analysis results can be used to construct predictive models for full surfaces. Especially, there are no "correct" answers in geostatistics, only the opportunity to gain more knowledge about the data and the measured surface.

The Spatial Dependence Modeler provides tools for measuring spatial variability or its complement, continuity in sample data. The module Model Fitting can construct models based on spatial variability with mathematical fitting techniques. The module Kriging and Simulation test models for the prediction and simulation of full surfaces.

5.1 TIN and TIN Surface

The TIN module generates a triangulated irregular network (TIN) model from either vector point or vector isoline input data. The triangulation is generated through a Delaunay triangulation that can be constrained or non-constrained, see Figure 5-1. The TIN generation window is shown as Figure 5-2.

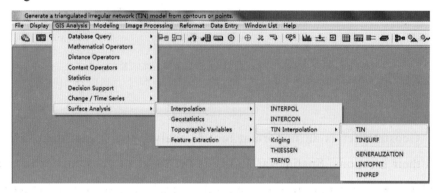

Fig.5-1　TIN operation in IDRISI

With the constrained option (reference to Figure 5-2), a contour vector file is required as input file; and the constrained option is provided to avoid triangle edges not cross isolines. With constrained option, some triangles may not meet Delaunay criteria; and the module TINPREP may be used to adjust the density of vertices along isolines in order to improve the triangulation.

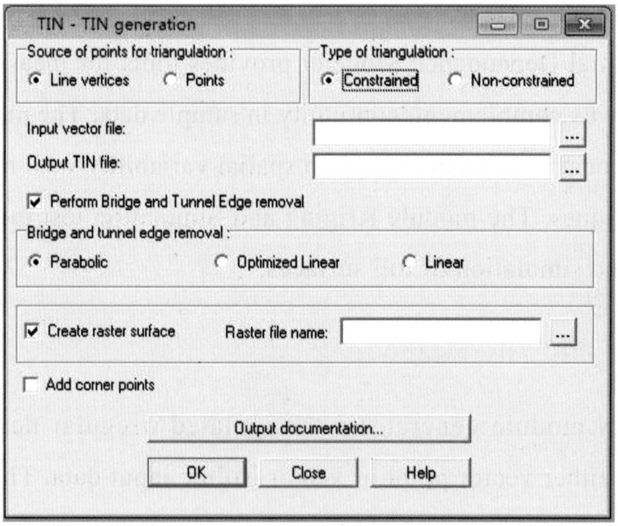

Fig.5-2　TIN window

If contour or isoline data were used to generate the TIN, the TIN can be optimized by the option of perform bridge and tunnel (B/T) edge removal. B/T edges are triangle edges that fall above or below the actual surface. In IDRISI, a TIN is created and all B/T edges are identified. New vertices are added at the midpoint of each B/T edge. These new vertices are then included in a re-triangulation of the network. The output of TIN is both a line vector file of the facet edges of the network and an ASCII .tin file. If the B/T edge removal option is selected, an output vector point file of the critical points is also produced. This is useful to extend the input vector data beyond the limits of the desired final study area when the TIN is to be used to create a raster image surface. This will ensure that the entire rectangular image contains data values rather than background values. An option to add corner points to the input vector data is provided to mitigate cases in which the desired raster surface is not entirely covered by the TIN. The module TINSURF may be launched automatically when click OK in TIN window, or run separately; thus an existing TIN file is used to generate a raster image surface. The TIN module is necessary for module TINSURF, which is shown as Figure 5-3. The TINSURF results will auto-display when you click ok in the TINSURF operation. Three methods are provided for interpolating attribute values for the added critical points. The recommended method is parabolic; of course, the optimized linear or linear interpolation method is optional. For a critical point, the attribute value is determined by interpolating the value for its position in up to eight directions. The final attribute assigned is the average of those interpolated for all valid directions.

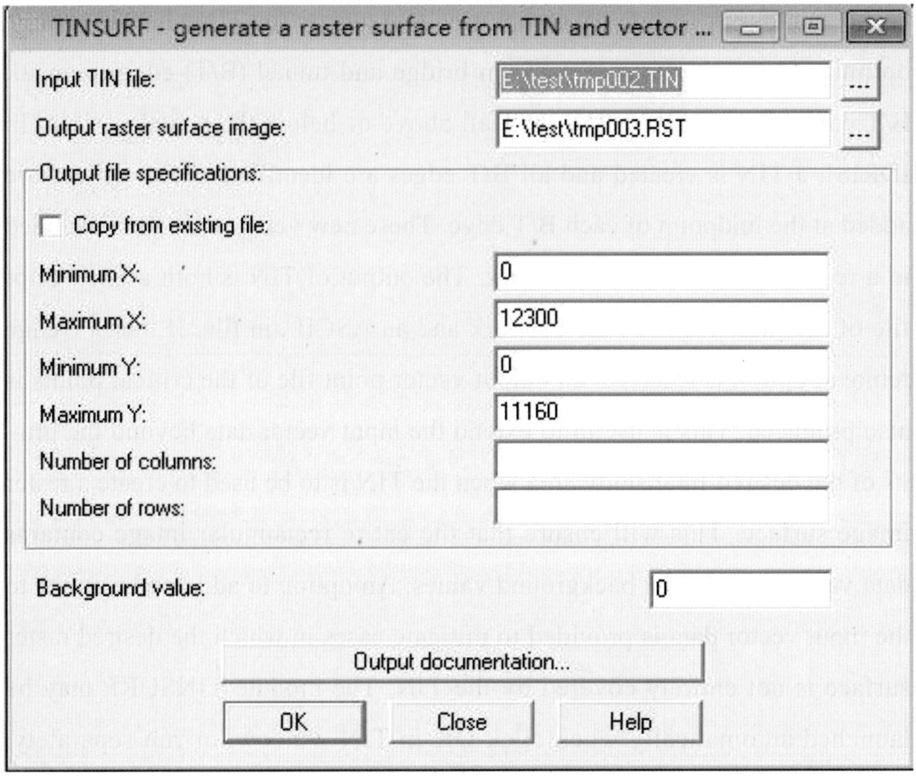

Fig.5-3   TINSURF module

If isoline data are used to create the TIN, the isolines should not cross. With the constrained option, an error message will be generated if an isoline intersection is detected. The input data should be corrected then TIN can be run again. The resulting TIN vector and ASCII files are given the same name, but have .vct/.vdc and .tin file name extensions respectively. If B/T edges were removed, a separate vector point file will be created with the name TINFILENAME_TIN_Critical_Points, where TINFILENAME is the name given to the output TIN file in the TIN dialog. The constrained triangulation is only available if the input data are isolines. B/T edge removal is also restricted to isoline input data. If a raster surface is to be created from the TIN, then the point or line IDs must be the attribute to be modeled. The first line of the ASCII TIN file (.tin) indicates whether the TIN was created with B/T edge removal or without B/T edge removal. If B/T edges were removed, the module TINSURF will automatically look for the critical point vector file that was generated by TIN. The name of the vector file used to create the TIN is stored in the TIN file. It is used by TINSURF to define the attributes of the vertices when creating a surface. Therefore this vector file should not be deleted nor renamed prior to running TINSURF.

If you wish to create a raster surface from the TIN, you should ideally use an input vector file that goes beyond the edges of the final raster image. This will ensure that the entire raster image has attribute values. If the input vector data you have does not extend beyond the edges of the raster extent, you may find that the resulting raster image has some areas around the edges that have been assigned the background value. The corner point option of the TIN dialog will help to alleviate this problem in the corners of the image.

TINSURF may be automatically launched from the TIN dialog box by selecting the output raster surface option. A raster surface image can be generated from a TIN model and the vector point or line file that defines the vertices of the TIN. The extent, number of columns, and number of rows of the output raster image can be defined in TINSURF module, references to Figure 5–3. For each raster pixel in the output image, an attribute value is calculated based on the positions and attributes of the three vertex points. Each pixel center will fall in only one TIN facet, but a single facet may contain several pixel center points. If we launch TINSURF dialog box from IDRISI menu, see Figure 5–3 and Figure 5–4.

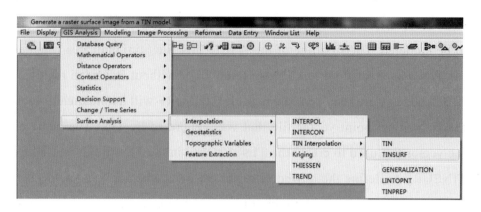

Fig.5-4　Launch TINSURF dialog box

In Figure 5-3, the name of the TIN file and the output raster surface image should to be selected or imputed. The min/max X/Y values can come from the TIN file default. The number of columns and rows are needed for the raster image. Also the min/max parameters may be modfied. The Output documentation button can be used to enter a title and value units, see Figure 5-5.

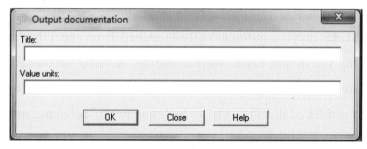

Fig.5-5　The Output documentation

Especially, the input vector file can be either integer or real ID type and must be file type binary. The number of output rows and columns must be integer values. The background value can be either an integer or real number. The TINSURF module is computationally intensive and thus will not run on images larger than 32 000 rows and columns.

5.2 Spatial Dependence Modeler

Spatial variability is assessed in terms of distance and direction. The spatial variability can be described as surfaces. This analysis is carried out on pairs of sample data points. Each pair is characterized by separation distance and direction. The distance is measured in units of lags, where the length of a lag is called lag distance or lag interval. For example, if the lag were defined as 10 kilometers, the third lag would include those data pairs with separation distances of 20 to 30 kilometers.

The semivariogram is a tool to describe spatial variability. For a data pair, the semivariogram is represented as,

$$\gamma(h) = \frac{1}{2n}\sum_{i=1}^{n}\left[z(x_i) - z(x_i + h)\right]^2 \qquad (5-1)$$

In which,

x_i, x_{i+h} = data pair;

n = number of pairs with distance h;

h = step length;

$z(x), z(x+h)$ = data value at (x_i, x_{i+h}).

The semivariogram can be presented as a surface image and a directional image. The Figure of surface image for semivariogram shows the average variability in all directions at different lags. The origin (center) position represents zero lags. The lags increase from the center toward the edges. The zero direction in the surface image is the direction from origin position to north, and right represents 90 degrees, and so on. The X-axis is distance (lags), while the Y-axis is the average variability of sample data pairs in every lag. The data for surface image can be restricted to any direction pairs, or regardless of direction.

Usually, the overall variability is analyzed firstly with regardless of direction that is treated as omnidirectional semivariogram in IDRISI. Then, in order to understand the structure of the data set, several plots may be worked out with different directions and lag distances. Four parameters are used to describe the structure of the data set; they are the sill, the range, the nugget and anisotropy. In most cases, spatial variability increases with the distance. The sill is a variance value when the curve reaches the plateau. The plateau means the point that the variability no more increases with separation distance between pairs. The range is the distance from the lowest variance to the sill. Sample data beyond this distance would not be considered in the interpolation process that will create risk or suitability images. The nugget is the variance at the distance of zero. Of course, zero is the desired value of the nugget. However, a non-zero value usually appears since the uncertainty of sample data that produce not spatially dependent variability. Anisotropy is the fourth parameter of the data set structure. There are two modes for spatial variability, isotropic model and anisotropic model. However, spatial variability is not isotropic in most cases; and it is called anisotropic model.

In many engineering problems, several influence factors should be considered, each can be represented as a surface image. The Spatial Dependence Modeler can be launched from the GIS Analysis/Surface Analysis/Geostatistics menu, see Figure 5-6 and Figure 5-7.

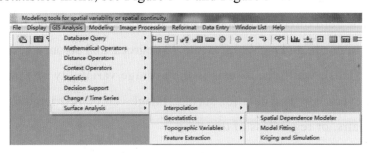

Fig.5-6　The Spatial Dependence Modeler

Chapter 5 Geostatistics

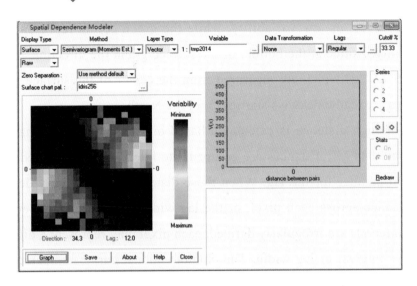

Fig.5-7   Dialog box of Spatial Dependence Modeler

Enter the input vector variable file and set the display type to surface. Accept the rest of the defaults, and then press the graph button. Once the variogram surface graph has been produced, the cursor can be used to check the direction and lag. The variogram surface is a representation of statistical space based on the variogram cloud.

The variogram cloud is the mapped outcome of variogram values for each resulting pair that each sample data point matches with every other sample data point. The results can be displayed according to their separation distance and separation direction. Superimposing a raster grid over the cloud and averaging cloud values per cell creates a raster variogram surface. Lag distance zero is located at the center of the grid, from which lag distances increase outwardly in all directions. Each pixel thus represents an approximate average of the pairs' semi-variances for the set of pair separation distances and directions represented by the pixel. When using the IDRISI standard palette, dark blue colors represent low variogram values, or low variability, while the green colors represent high variability. Notice at the bottom of the surface graph the direction and the number of lags which are measured from the center of the surface graph. Degrees are read clockwise starting from the North.

Although distance is calculated based on the spatial coordinates of the input data set, distances are grouped into intervals and assigned a sequential number for the lag. When those distances are regularly defined, the distance across each pixel, or the lag width, is the same. When distance intervals are irregularly defined, each pixel may represent a different distance interval or lag width. The omnidirectional curve summarizes the surface graph on the left by plotting for each lag the average variability of all data pairs in that lag.

5.3 Model Fitting

The model fitting describes the pattern of spatial variability of the measured surface as continuous model. The experimental semivariogram suggest the model's form. Those semivariogram that resulted from the Spatial Dependence Modeler module are used to design models. First, mathematical curves are designed as a proposed model variogram, and automatic methods are used to refine the fit. Mathematical fitting can inversely weight sample semivariogram lags by the number of pairs that were averaged when the semivariance was calculated. However, designing a curve is best done as both a visual and an automatic process. To facilitate the Model Fitting module, an isotropic model or an anisotropic model should be designed.

To do an isotropic analysis, two omnidirectional variogram models should be created with the Spatial Dependence Modeler. Model Fitting module can be launched from GIS Analysis/Surface Analysis/ Geostatistics menu, references to Figure 5-6. The parameters for the structure(s) will describe the mathematical curves that constitute a model variogram, see Figure 5-8.

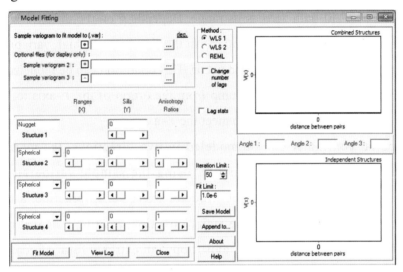

Fig.5-8   Dialog box of Model Fitting module

These parameters include the *sill, range*, and *anisotropy ratio* for each structure. When there is no anisotropy, the anisotropy ratio is represented mathematically as a value of 1. The *sill* in Model Fitting is an estimated semivariance that marks where a mathematical plateau begins. The plateau represents the semivariance at which an increase in separation distance between pairs no longer has a corresponding increase in the variability between them. Theoretically, the plateau infinitely continues showing no evidence of spatial dependence between samples at this and subsequent distances. It is the semivariance where the range is reached.

In designing a model to fit to the sample data, the general shape of the curve is defined by the mathematical model(s) that are used. In the Model Fitting interface, the first structure of the model listed is the Nugget structure. The Nugget structure does not affect the shape of a curve, only its Y-intercept. It has been listed separately from other structures because many environmental data sets experience a rise in the Y-intercept for the curve. Graphically it appears as a sill with zero as its range. Depending on the distance interval used, and the number of pairs captured in the first interval, high variability at very close separation distances can occur. This condition can be modeled with a Nugget structure which is the jump from the origin of the Y-axis to where the plot of points appears likely to meet the Y-axis.

Spatial continuity can be modeled in two directions, the maximum continuity and minimum continuity, using the sample variogram files saved in the Spatial Dependence Modeler exercise. In practice, a model must be built firstly based on the major direction, treating it as if it represented an isotropic model. Zonal anisotropy, an extreme form of geometric anisotropy, occurs when there is a noticeable difference in the degree of variability across distance in the direction of maximum variability relative the direction of minimum variability. To model the zonal effect, we specify an ellipse in the direction of maximum variability. To build a model containing zonal anisotropy, the first structure is the zonal structure which will be given a very low anisotropy ratio. The model variogram can be used to develop kriging or simulation surfaces.

5.4 Ordinary Kriging

In kriging interpolation processes, directional semivariogram are used to create values of unknown points. A smooth curve of spatial variability that represented by directional semivariogram is needed. The smooth curve is presented as mathematical function. Sills, ranges, nugget and anisotropies structures are defined for the smooth curve. Then a surface can be interpolated with the Kriging model. The model will define weights of sample data those are combined to produce values for unknown points. The weights associated with sample points are determined by direction and distance to other known points.

For one unknown point x, the value can be calculated as,

$$z_x^* = \sum_{i=1}^{n} \lambda_i z(x_i) \quad i=1,2,3,\ldots,n \qquad (5-2)$$

In which,

z_x^* =value of unknown point;

λ_i =weight for $z(x_i)$;

$z(x_i)$ =value at known point x_i.

For kriging interpolation,

$$\sum_{i=1}^{n} \lambda_i = 1 \quad i=1,2,3,\ldots,n \qquad (5-3)$$

In this part, the parameters of the variogram model are used to create an interpolated surface with ordinary kriging. Since the minimized variance error, ordinary kriging is known as a Best Linear Unbiased Estimator (B.L.U.E.). Two images will be created as surfaces of kriged estimates and estimated variances.

Kriging estimates a new attribute for each pixel on the basis of a local neighborhood. A quick method for evaluating the fit of a model variogram is through cross validation. With cross validation, the algorithm interactively removes one sample and interpolates a new value for the sample data location based on the input model and other input parameters. It continues this procedure for each data sample until all sample data locations have an estimated value. The end result is a new image with interpolated points only at the original data points, and another image containing variances. A table comparing the original data values to the new data values with their related statistics is also created.

From the GIS Analysis/Surface Analysis/Geostatistics menu, choose Kriging and Simulation. Select Ordinary Kriging and the cross validation option under Kriging Options. Enter RAINPRED as the model source, and then click on the edit option under Model Specifications, see Figure 5-9. Then, enter the input sample vector data file, select the maximum number of sample points under the local neighborhood selection options. A mask file is needed that specifies the rows, columns, and reference system of the area to be predicted. Enter the mask image file and the output Prediction File. Finally, enter the output Variance File. Press OK and examine the module results when cross validation finishes.

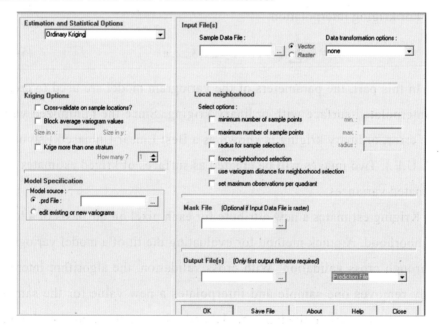

Fig.5-9　Dialog box of Ordinary Kriging

The module results show the correspondence between the original and the predicted values. The sample data and the predicted values should be fairly similar. Different sample data can be used, such as fewer samples, radius, or a radius with a quadrant search of a few samples per quadrant. With a combination of cross-validation, different selection of the local neighborhood options, and the editing of the parameter file, the prediction process can be improved.

Interpreting a variance image is not straight forward. It never confirms the correctness of the chosen model, but only provides evidence of problems. A different variance image based on a different model could show different uniformity in the variances across all areas. Cross validation and variance surfaces can be used to evaluate inconsistencies, ultimately to decide whether to modify or to reject the model, but not to evaluate prediction accuracy. As an investigative manner, there are many ways to use geostatistical tools in IDRISI.

Chapter 6　Application Examples of Geostatistics

In this chapter, several application examples of geostatistics will be introduced.

6.1 Application in Air Pollution

In this application, 16 cities' air pollution data are used to analyze the distribution and influence factors. The air pollution data come from the website http://www.cnemc.cn/publish/totalWebSite/0666/newList_1.html, they are PM2.5 data from January 2015 to May 2015 of 16 cities in Jiangsu Area. The data is shown in Table 6-1.

Table 6-1　Air pollution data of PM2.5

City	X-axis	Y-axis	May	April	March	February	January
Lianyungang	1.4	4	40	42	43	77	91
Yancheng	4.8	4.2	31	42	45	72	84
Huaian	3.2	2.6	43	50	60	88	94
Nantong	8.2	4	42	56	60	71	96
Suqian	1.8	1.8	45	56	69	79	90
Changzhou	7.7	2.2	46	61	66	79	109
Suzhou	9.3	2.7	46	63	61	72	97
Wuxi	8.5	2.5	49	59	60	74	101
Zhenjiang	6.4	1.7	49	52	57	68	88

Continued

City	X-axis	Y-axis	May	April	March	February	January
Nanjing	6.1	0.5	52	51	56	73	96
Taizhou	6.4	2.8	52	60	61	74	98
Yangzhou	6	2	54	50	69	65	81
Xuzhou	0.2	0.2	65	57	74	90	101
Shanghai	10.2	4.1	40	55	54	63	81
Jiaxing	10.5	2.4	44	52	50	62	87
Huzhou	9.7	1.4	46	54	58	70	101

To generate evaluation images with the data in Table 6-1, database workshop need to be launch firstly. There are several methods to launch database workshop, and we start it from GIS Analysis menu. The procedure is begin from GIS Analysi → database query to start database workshop, see Figure 6-1. In the database workshop window, in File menu click New to establish a new mdb file and input new mdb file name. In database workshop, from file menu, choose Import → table → from external file, click from external file to import, click open in Figure 6-2, and in the pop up windows select sheet 1, then click OK, and click Yes in the pop up window, this procedure can reference to application example in 2.3.2. The data is import as Figure 6-3.

Chapter 6　Application Examples of Geostatistics

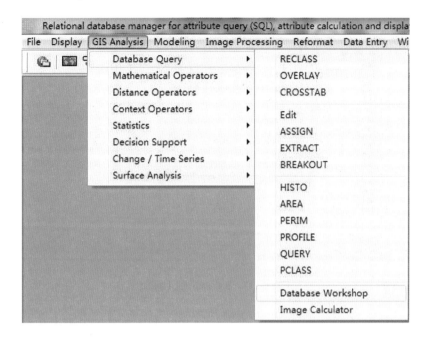

Fig.6-1　Launch database workshop

Fig.6-2　Select input data

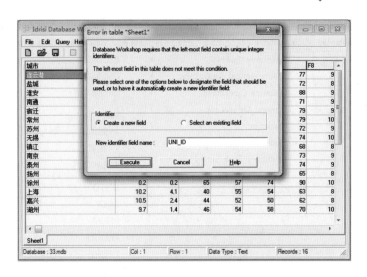

Fig.6-3　Select input data

In Figure 6-3, click "Execute", the input data is shown as Figure 6-4. In Figure 6-4, click right mouse button and select remove field, see Figure 6-4. In the pop up window (Figure 6-5), click OK, the input data shown as Figure 6-6.

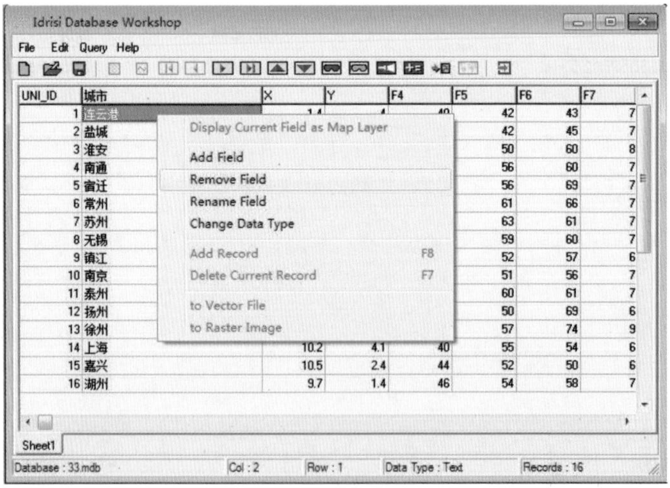

Fig.6-4　Input data

Chapter 6   Application Examples of Geostatistics

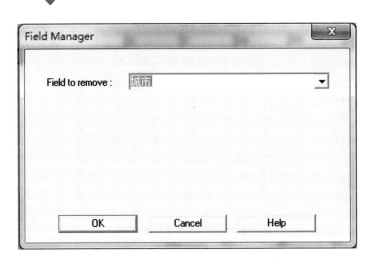

Fig.6-5   Remove field

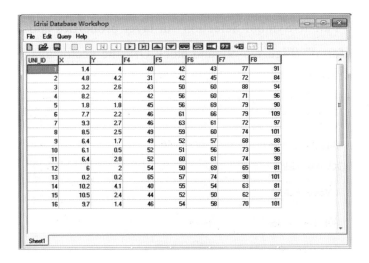

Fig.6-6   The application example data

Then click Save and choose File → Export → Field → X,Y to Point Vector File in file menu as Figure 6-7.

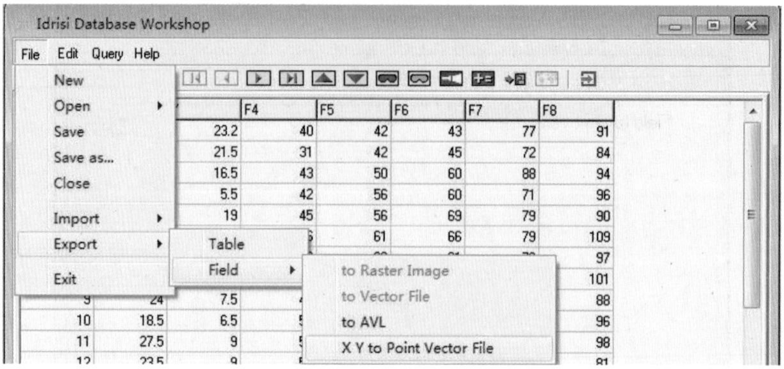

Fig.6-7 Export x, y data

Choose X-axis in x values field, and Y-axis in y values field, and then click OK in export vector file in Figure 6-8. In the vector image, select "Advanced Palette/Symbol Selection" in layer properties and then select "None (uniform)" (Figure 6-9), and clock OK, the vector image is shown as Figure 6-10. This procedure can reference to 2.3.2.

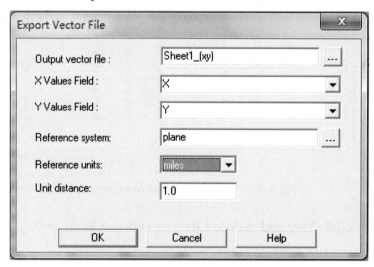

Fig.6-8 Export vector file

Chapter 6   Application Examples of Geostatistics

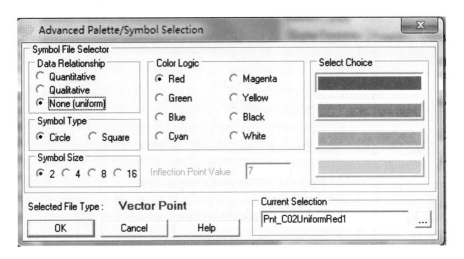

Fig.6-9   Advance palette/symbol selection

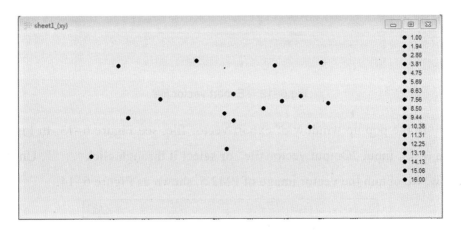

Fig.6-10   The x,y vector image

Click the button of "establish a display link" to establish between (x,y) vector file and input data file, it is shown as Figure 6-11. Please pay attention to Figure 6-11, the link field name must choose "UNI_ID". In database workshop, choose File → Export → Field → to Vector File, see Figure 6-12.

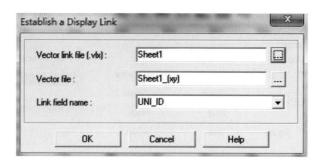

Fig.6-11　Establish a display link

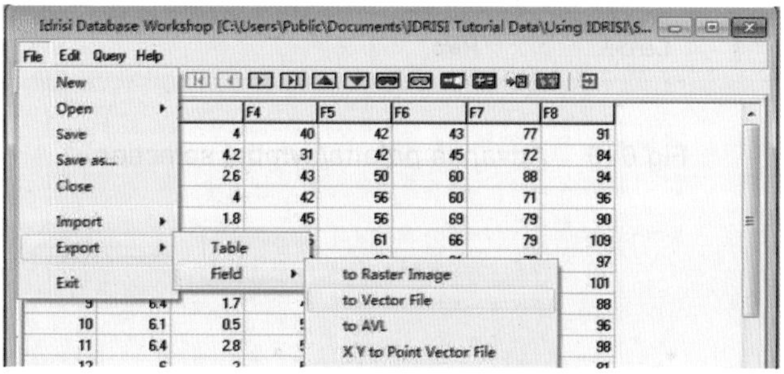

Fig.6-12　Export vector file

In the pop up window of export vector file, see Figure 6-13. In Figure 6-13, input "Output vector file" or select it through click " … ". Until now, we obtain the vector image of PM2.5, shown as Figure 6-14.

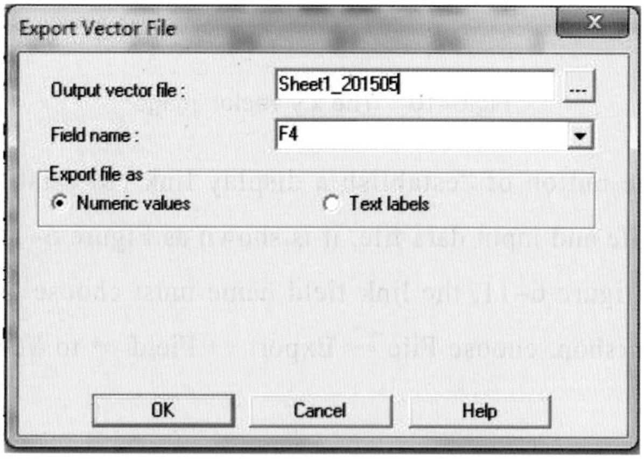

Fig.6-13　Export vector file window

Chapter 6　Application Examples of Geostatistics

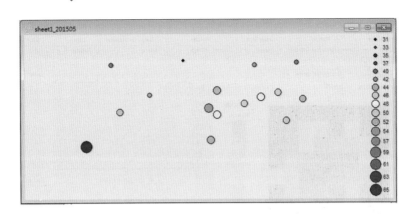

Fig.6-14　The vector file of PM2.5

First, we create a raster image of air pollution data with TIN method, see Figure 6-15. The procedure can reference to 2.3.2.

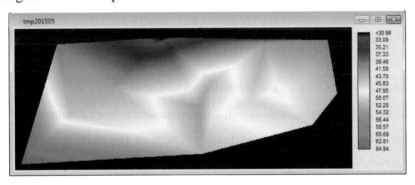

Fig.6-15　The vector file of PM2.5

Now we begin to use the Spatial Dependence Modeler. Open the Spatial Dependence Modeler from the GIS Analysis/Surface Analysis/ Geostatistics menu. Select "sheet1_201505" as the input vector variable file, see Figure 6-16. Choose surface as the Display Type, and press the graph button to create the variogram surface.

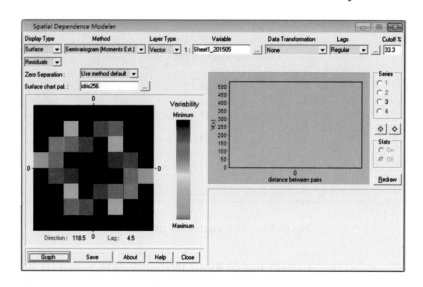

Fig.6-16　Spatial Dependence Modeler

In Figure 6-16, a variogram value for each resulting pair is created. Variogram values are displayed by separation distance and separation direction. In this example, the default method was used, lag distance zero is located at the center, and lag distances increase from the center. Thus, each pixel represents an approximate average of the pairs' semivariance. Low variability is represented by dark blue colors, and high variability is represented by green colors. Please notice that degrees are read clockwise starting from the North, the direction and the numbers of lags are graphed at the bottom of the surface. But in this image, 0 degree represents the Northeast, not the North.

Back to the Spatial Dependence Modeler dialog, in the lags box, click on the small button to the right to access the regular intervals lag specification dialog. The default number of lags is 10, and the lag width is calculated automatically. Also, you can try to change to manual mode, and leave the lag width to default value. You can change the Cutoff % value,

Chapter 6 Application Examples of Geostatistics

and new image can be created by press the Graph button. By specifying 100, Stats will calculate semivariance for all data pairs, overriding the specified number of lags and lag width. From this image, The direction of Northeast to Southeast has the maximum variability. And a low variability pattern is in the east–west or north–south direction.

Figure 6-17 is the variability pattern with directional variograms. This is fulfilled through change the display type parameter from surface to directional, and change residuals to raw. Check the omnidirectional override in the lower-right section of the dialog box. The image is fulfilled by press Graph button.

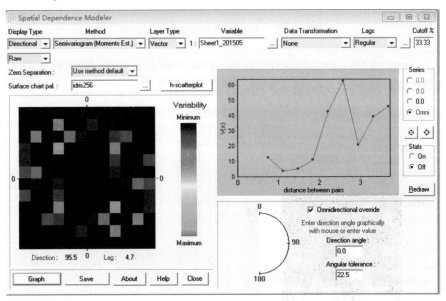

Fig.6-17 Direction type of spatial dependence modeler

In Figure 6-17, each point summarizes the variability calculated for data pairs that regardless of the direction. Figure 6-18 is the result of directional variograms when you choose stats on.

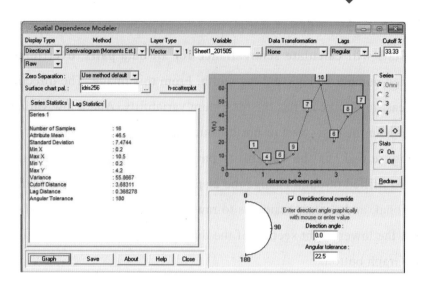

Fig.6-18　Stats on in direction type

Uncheck the omnidirectional override, and input the degree of direction angle, you can obtain the min or max variability direction.

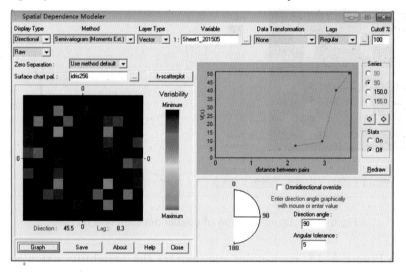

Fig.6-19　Stats on in direction type

Click on Save button to save variogram file, such as omnidirection, min or max direction.

Chapter 6 Application Examples of Geostatistics

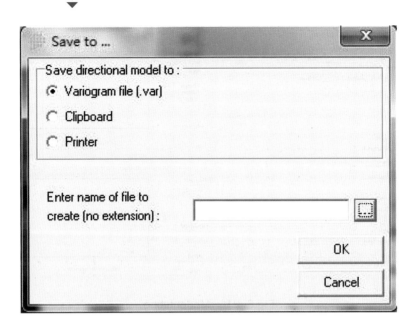

Fig.6-20   Stats on in direction type

The directory information is stored in .var files from which they were created. Now we begin model fitting module. Enter omni file as the Sample Variogram model to fit, and you can also choose the second and third sample variogram files, such as min and max direction variogram file, see Figure 6-21. With model fitting, the continuity structures are interpreted by the semivariogram produced with the Spatial Dependence Modeler. The parameters for the structure(s) will describe the mathematical curves that constitute a model variogram. These parameters include the sill, range, and anisotropy ratio for each structure. When there is no anisotropy, the anisotropy ratio is represented mathematically as a value of 1. The sill in Model Fitting is an estimated semivariance. The plateau represents the semivariance at which an increase in separation distances. For the Structure 2, exponential model is used and enter a Range of 3 and a Sill of 55 in the set of corresponding boxes.

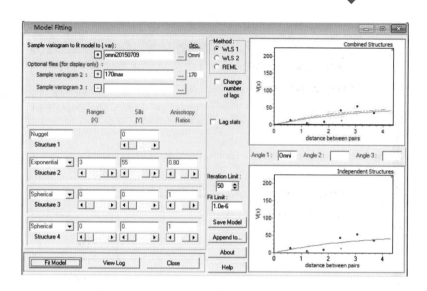

Fig.6-21　Model fitting

When you finish this, click Fit Model, and then click Save Model. Input the output file, it is shown as Figure 6-22.

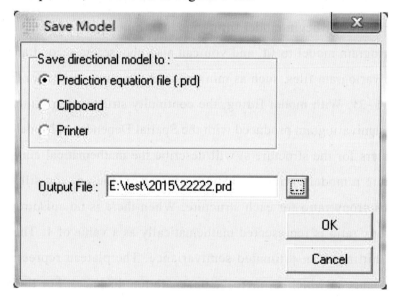

Fig.6-22　Model fitting

Chapter 6 Application Examples of Geostatistics

In designing a model to fit to the sample data, the general shape of the curve is defined by the mathematical model(s) that are used.

The last step is to launch Kriging and Simulation module from the GIS Analysis/Surface Analysis/Geostatistics menu.

Choose Ordinary Kriging as the default estimation option and enter original sample vector file as input data file. Choose 22222.prd as the model source under Model Specifications. In Kriging and Simulation module, the maximum number of sample points or radius of sample selection must be selected. In this application, 6 as radius of sample selection or 9 as the maximum number of sample points is inputted, see Figure 6-23. Please notice that a mask file must input to specify the rows, columns, and reference system of the area to be predicted. The mask file is provided as Figure 6-24. Of course, you can try to create a mask file by yourself.

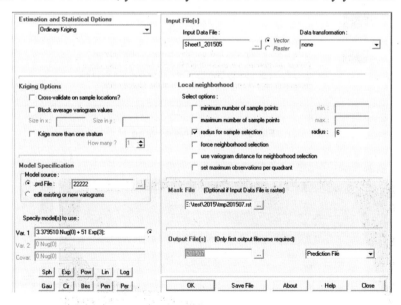

Fig.6-23 Stats on in direction type

Fig.6-24   Mask file

Enter 201507 as the output Variance File, and press OK, a new image with ordinary kriging will be created. During this, analysis stage will be displayed as Figure 6-25. The calculate result is shown as Figure 6-26.

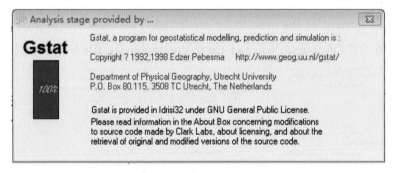

Fig.6-25   Analysis stage

Fig.6-26   Result of ordinary kriging interpolation

Chapter 6 Application Examples of Geostatistics

In order to display the location of cities, we can use "add layer" function to add cities location, see Figure 6-27. The add layer interface is shown as Figure 6-28.

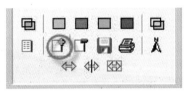

Fig.6-27   Add cities location vector layer

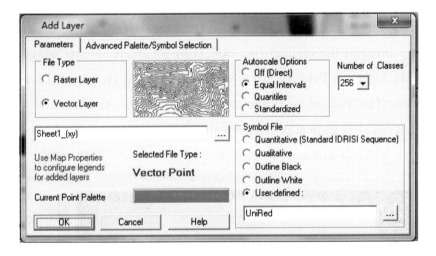

Fig.6-28   Add layer interface

The final module result is shown as Figure 6-29. We can analyze the difference with different method through the correspondence between Figure 6-15 and Figure 6-26 or Figure 6-29. From the comparison, we can find that a more perfect result will be obtained with geostatistics method, and the core is Kriging estimation method.

The result can be adjusted, and many tools can be used to enhance the display.

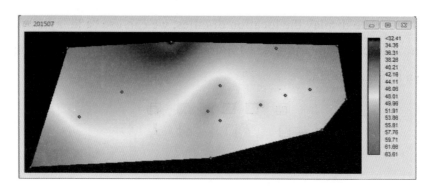

Fig.6-29　Final result with cities location

Figure 6-29 is the result of PM2.5 from the data of may 2015. We use this to represents the summer situation. Now we create another image with the PM2.5 data in Febrary 2015 (see Table 6-1) to represent the winter situation. This procedure is same with above steps. And the TIN interpolation result is shown as Figure 6-30. The result of spatial dependence modeler is shown as Figure 6-31.

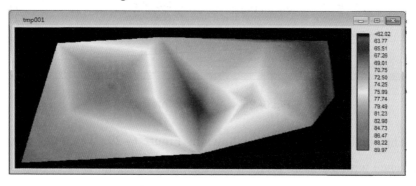

Fig.6-30　Result of Feb. 2015

Chapter 6 Application Examples of Geostatistics

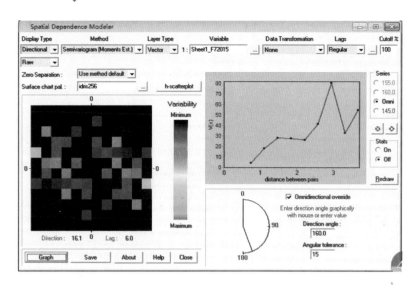

Fig.6-31   Spatial dependence modeler result

We can do all the geostatistics procedure as above, the second step is to do model fitting as Figure 6-32.

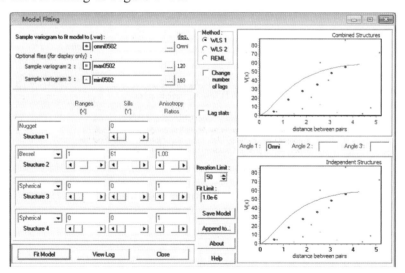

Fig.6-32   Model fitting (2)

And the third step is estimation and statistical options, which is shown

147

as Figure 6-33 and Figure 6-34.

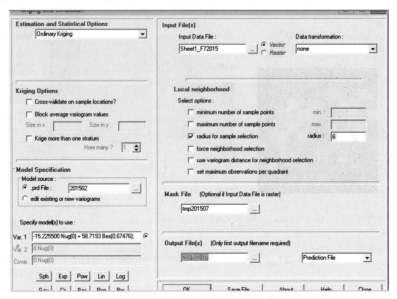

Fig.6-33　The third step

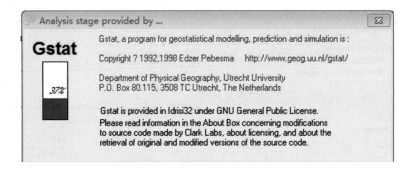

Fig.6-34　Analysis stage (2)

The final result is shown as Figure 6-35.

Chapter 6　Application Examples of Geostatistics

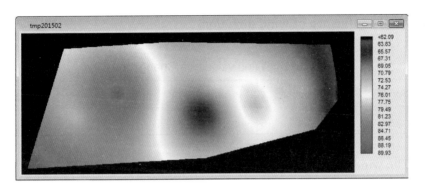

Fig.6-35　The final result with data of Feb. 2015

We can analyze the Figure 6-35 and Figure 6-29, which represent the PM2.5 distribution in summer and winter. It can be found that there is more air pollution in winter, this can be certified by the air pollution index in 2013 in Changzhou City, see Figure 6-36.

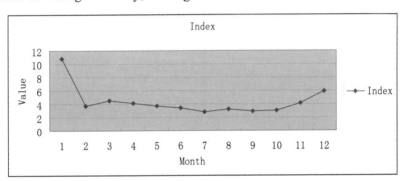

Fig.6-36　Air pollution index of Changzhou in 2014

From all above, we can find that air pollution is more serious in winter. Now, we analyze the reasons for this.

The wind velocity in Changzhou City is shown in Table 6-2, it can be found that wind velocity is not the main reason for air pollution distribution. The min wind velocity is not the lightest air pollution season, and the max wind velocity is also not the heaviest air pollution season. Then we analyze the influence of wind direction.

Table 6-2 Wind velocity in Changzhou City

Spring	Summer	Autumn	Winter	Average
3.14	2.92	2.63	2.75	2.86

The wind direction is shown as Figure 6-37. In Figure 6-37, the wind direction in winter is represented by dotted line, and the wind direction in summer is represented by solid line.

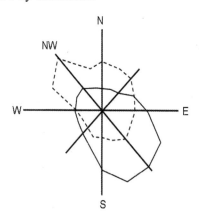

Fig.6-37　Wind direction

From Figure 6-37, it can be found that the main wind direction in winter is northwest. Because there are more serious air pollution in that direction, thus more serious air pollution happened in winter. During summer, the main wind direction is southeast where is the sea and no air pollution there. Above all, air pollution is affected by wind direction.

6.2 Application in Casing Failure Analysis

As an example application, five factors in J607 block of Liaohe Oilfield are selected. They are average pressure, max temperature, volume and times

Chapter 6　Application Examples of Geostatistics

of steam injection, and volume oil production. The data of five factors in 119 oil wells in J607 Block of Liaohe Oilfield are shown in Table 6-3. The purpose of this study is to analyze the spatial variability, and then develop surface maps based on the sample data. The resultant surface maps are used to evaluate the risk of oil production.

Table 6-3　Data in J607 Block

Wells	X-axis	Y-axis	Times of Inj.	Volume of Inj.	Volume of production	Max temperature	Average pressure
44-74	147	139	12	26 520	38 253	338	12.525
44-76	140	131	11	30 007	34 015	340	12.7
44-78	132	124.5	8	16 797	21 368	341	11.91
44-80	125	115	12	26 672	30 291.5	343	13.84
44-82	121	108	8	14 346	10 545	355	14.56
44-84	110	100.5	6	12 451	12 267	350	14.83
46-68	188	155	7	13 787	8 970	346	15.73
46-70	168	145	12	19 576	24 437.8	350	15.79
46-74	154	132	11	27 895	37 101	345	12.72
46-76	147	123	12	21 381	41 022	338	11.68
46-78	141	116	17	25 302	34 033	333	11.58
46-80	133	110	8	14 734	28 380	343	12.53
46-82	124	101	7	14 203	15 330	355	15.91
46-84	114	92	8	14 448	25 255	348	16.2
46-88	99	80	8	13 279	30 264	343	13.4
46-721	169	139	9	20 470	16 574	330	9.94
46-741	163	131	9	14 619	28 477	348	13.68
46-761	154	125	8	16 804	19 708	349	12.38

151

Continued

Wells	X-axis	Y-axis	Times of Inj.	Volume of Inj.	Volume of production	Max temperature	Average pressure
48-66	198	155	17	35 453	50 579	338	12.89
48-68	191	148	12	20 920	34 308	325	10.56
48-70	179	139	13	29 883	47 441.1	348	10.94
48-72	170	131	13	29 359	45 070	347	11.49
48-74	163	123	2	2 445	1 680	348	15.45
48-76	156	118	12	21 574	38 851	346	12.57
48-78	148	110	8	7 338.7	21 764	335	11.28
48-80	140	101	11	19 020	37 101	338	12.73
48-82	134	93	11	15 826	37 101	361	15.46
48-84	125	85	7	15 878	24 760	343	14.41
48-86	115	80	9	15 514	30 298	355	14.59
48-721	181	131	10	14 798	20 017	341	11.15
48-741	172	124	7	12 642	21 572	337	10.36
48-761	164	116	8	19 736	26 444	348	14.2
48-K74	165	125	8	14 252	14 454	341	11.26
49-86	118	73	6	8 305	11 711	347	15.57
50-66	204	148	10	30 011	42 103.5	331	10.27
50-68	197	138	12	30 384	45 723	335	11.26
50-70	190	131	18	34 113	45 119	339	10.28
50-72	182	123	14	26 133	29 375	345	10.74
50-74	170	116	12	27 670	39 078	339	12.26
50-76	162	108	13	30 412	35 656	335	11.37
50-78	156	100	8	13 346	17 379	349	12.83

Chapter 6 Application Examples of Geostatistics

Continued

Wells	X–axis	Y–axis	Times of Inj.	Volume of Inj.	Volume of production	Max temperature	Average pressure
50–80	148	92	6	9 441	37 097	350	12.25
50–82	141	85	4	6 563	21 962	345	13.75
50–84	133	79	5	9 342	27 435	348	14.16
50–86	128	71	9	16 170	13 184	353	13.09
50–661	212	145	6	15 521	13 383	332	10.6
50–701	200	132	12	27 148	33 795	341	12.98
50–721	188	122	7	14 099	21 977	325	12.27
50–741	180	116	7	16 805	19 749	346	15.01
52–64	219	148	10	19 953	67 598	326	11
52–66	208	138	14	27 269	38 399	334	10.76
52–68	204	131	13	27 589	42 600	326	11.36
52–70	197	125	16	33 765	18 386	337	11.28
52–72	188	115	13	26 701	43 321	338	10.78
52–74	189	109	12	23 168	49 565	338	13.02
52–76	172	100	14	27 731	39 710	340	11.81
52–78	164	92	9	12 577	27 923	340	12.6
52–80	155	85	8	12 984	21 629	340	14.66
52–82	148	77	7	15 136	27 221	349	14.2
52–84	140	71	3	5 693	35 167	352	15.07
52–86	133	62	9	15 127	17 652	342	14.5
52–641	227	147	4	8 131	28 312	326	10.95
52–661	219	140	9	18 610	34 785	324	10.29
52–681	212	132	8	13 311	31 625	331	11.56

153

Continued

Wells	X-axis	Y-axis	Times of Inj.	Volume of Inj.	Volume of production	Max temperature	Average pressure
52-701	204	126	9	18 694	28 158.5	333	10.58
52-721	197	114	7	12 153	29 852	321	10.31
52-741	181	108	8	16 177	27 424.7	343	13.05
54-66	221	132	17	37 270	36 730	342	10.82
54-68	211	124	16	30 662	30 590.9	329	11.08
54-70	203	118	15	53 255	45 458.1	340	11.92
54-72	195	108	10	20 651	38 303	326	10.48
54-74	189	102	17	31 419	31 605	347	11.86
54-76	180	93	10	22 516	36 927	333	11.65
54-78	171	85	7	11 361	21 320	338	12.1
54-80	162	78	6	11 343	19 491	358	13.02
54-82	151	70	3	3 988	9 415	341	14.57
54-84	146	62	4	5 114	42 314	342	14.28
54-621	242	147	6	12 620	22 661	320	10.92
54-641	236	141	8	13 794	28 796	328	11.04
54-661	225	132.5	7	12 696	18 464	344	11.69
54-681	219	126	6	17 194	24 411	328	11.1
54-701	213	120	7	11 820	24 300	343	12.76
54-721	204	109	5	13 145	21 364	346	14.06
54-X60	227	141	14	36 704	56 825	326	10.28
55-82	162	68	5	8 167	14 338	336	11.68
56-62	241	138	10	18 626	60 278	333	11.5
56-64	234	131	12	20 883	63 916	350	10.76

Continued

Wells	X-axis	Y-axis	Times of Inj.	Volume of Inj.	Volume of production	Max temperature	Average pressure
56–66	224	125	13	32 596	61 507	346	11.53
56–70	211	108	14	29 964	44 474.8	337	10.46
56–72	205	102	12	21 237	43 523	342	11.31
56–74	196	94	11	17 454	39 031	328	10.43
56–76	188	85	13	24 565	37 567	397	12.2
56–78	179	79	9	20 010	24 964	344	12.38
56–80	171	71	6	12 246	14 215	340	12.95
56–621	249	140	7	10 723	21 742	340	12.03
56–641	242	135	8	13 794	28 796	328	11.04
56–661	235	125	7	13 768	34 200	342	11.71
56–681	227	119	9	17 463	37 593	328	10.6
58–62	248	131	16	33 387	58 551	350	11.17
58–64	240	125	11	19 381	68 697	339	10.49
58–66	233	118	14	24 912	54 992	323	10.44
58–68	226	109	14	17 442	46 233	345	10.91
58–70	218	102	14	23 945	41 133	326	10.71
58–72	210	92	13	25 299	47 628	344	11.32
58–74	202	86	12	24 139	39 673	327	10.52
58–621	260	135	6	8 998	25 098	346	12.22
58–641	250	125	5	6 489	49 172	332	11.9
58–661	244	116	8	12 927	29 057	345	11.65
58–681	234	109	4	6 660	37 077	334	12.33
60–62	258	124	11	16 640	47 176	342	11.95

Continued

Wells	X-axis	Y-axis	Times of Inj.	Volume of Inj.	Volume of production	Max temperature	Average pressure
60-64	245	116	11	16 402	49 208	322	11.66
60-66	242	109	9	12 663	31 975	328	12.04
60-68	234	101	13	38 435	39 125	348	11.22
60-70	225	94	11	23 584	41 891	356	12.99
60-641	258	116	4	5 734	10 044	341	12.53
60-661	251	108	6	8 456	6 718	345	11.72
J607	221	118	16	40 475	36 153	341	11.47
J608C	109	87	4	8 089	3 764	339	14.6

According to the method above, spatial variability is analyzed, model fitting is fulfilled and ordinary kriging is use to create interpolating surfaces. The data of each factor in 119 wells are entered as the input vector variable file. The semi-variogram surface maps of five factors are calculated, which is shown as Figure 6-38 to Figure 6-42.

In semi-variogram surface images, lag distance zero located at the center of the grid, from which lag distances increase outwardly in all directions. Each pixel thus represents an approximate average of the pairs' semivariance for the set of pair separation distances and directions. As colors from dark blue to green colors, variability increase from low to high. The direction and the number of lags are displayed in the bottom of the semi-variogram surface image. From Figure 6-38 to Figure 6-42, we can analyze the spatial variability of five influence factors.

Chapter 6 Application Examples of Geostatistics

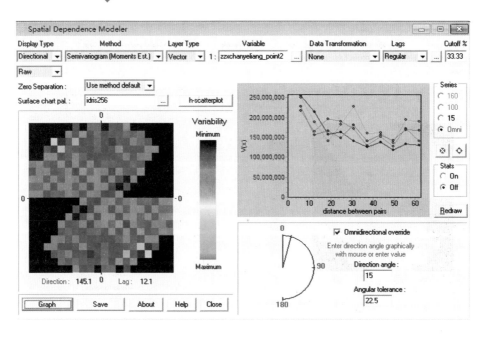

Fig.6-38 Semi-variogram of volume of production

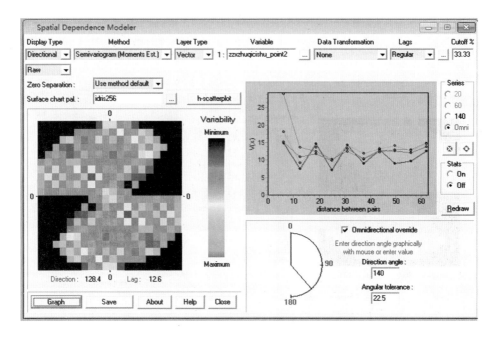

Fig.6-39 Semi-variogram of times of injection

基于GIS的决策支持与表面分析
GIS-Based Decision Support & Surface Analysis

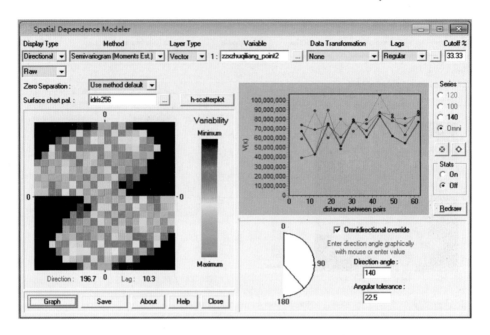

Fig.6-40　Semi-variogram of volume of injection

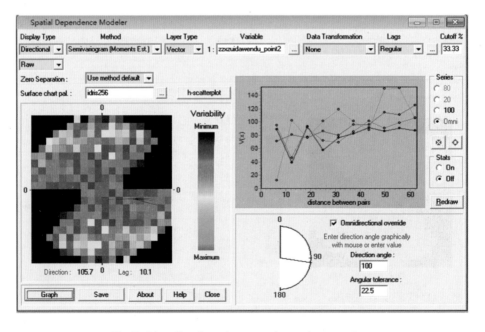

Fig.6-41　Semi-variogram of max temperature

Chapter 6 Application Examples of Geostatistics

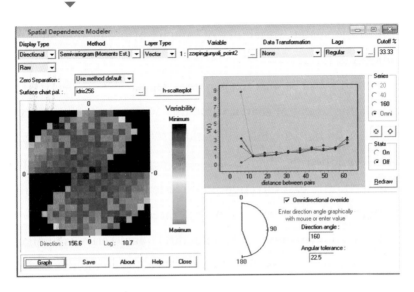

Fig.6-42 Semi-variogram of average pressure

Figure 6-38 is the spatial variability of volume of production. It can be found that spatial variability of volume of production is low in total. The max direction of spatial variability is 100 degree, but the difference is not very much. This can be certified by the direction series, and can be found that the variability is low in volume production.

Figure 6-39 is the spatial variability of times of steam injection. The max direction of spatial variability is 60 degree, and the min direction is 140 degree. The omnidirectional series is similar to an average in all directions and therefore it falls between the two series of 60 and 140 degree. The spatial variability of times of steam injection is high than the spatial variability of volume of production. This can be certified by the direction series (reference to the right image of Figure 6-39).

Figure 6-40 is the spatial variability of volume of steam injection. The max direction of spatial variability is 100 degree, and the min direction is 140 degree. The spatial variability of volume of steam injection is high than the spatial variability of volume of production and times of steam injection.

The fourth factor is max temperature. The 20 degree and 100 degree series reveal the extent of spatial variability with anisotropy and trending. The minimum spatial variability direction (100 degree) and the maximum (20 degree) implied the degree of spatial dependency across distance is greater in the north-east direction. In direction 20 degree, variability increases rapidly. In the nearly orthogonal direction at 100 degree, variability increases much slowly. The omnidirectional series is similar to an average in all directions and therefore it falls between the two series of 20 degree and 100 degree.

The fifth factor is average pressure of steam injection, and spatial variability of average pressure is shown as Figure 6-42. It is similar with the first factor, volume of oil production. It implies that the influence of total oil production is similar with average pressure of steam injection. Also, volume of injection and times of steam injection is similar. These reveal that steam injection times and volume of steam injection are controlled by geological condition; average injection pressure and total oil production are affected by production.

With the sample data of five influence factors as the input vector data files, interpolating surfaces of all five influence factors are obtained and shown as Figure 6-43 to Figure 6-47. In those images, the risk increase from low to high, therefore, they can be standardized with fuzzy to 0 to 255. 0 represents the minimum risk, and 255 represent the maximum risk. Then, the risk of casing failure can be evaluated.

Chapter 6 Application Examples of Geostatistics

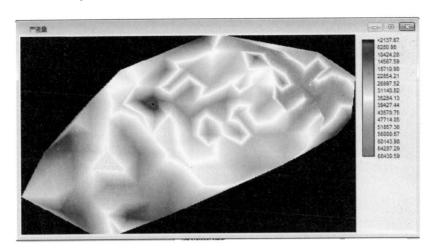

Fig.6-43 Interpolating surface of production volume

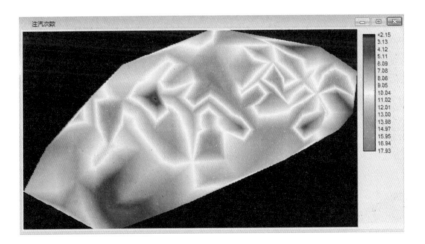

Fig.6-44 Interpolating surface of injection times

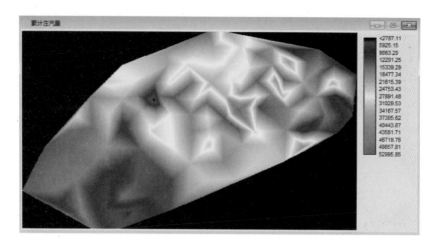

Fig.6-45　Interpolating surface of injection volume

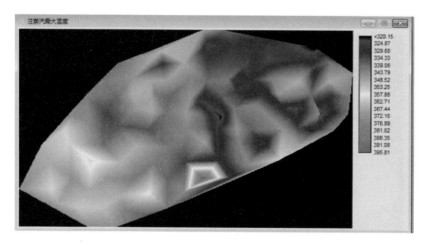

Fig.6-46　Interpolating surface of max temperature

Chapter 6 Application Examples of Geostatistics

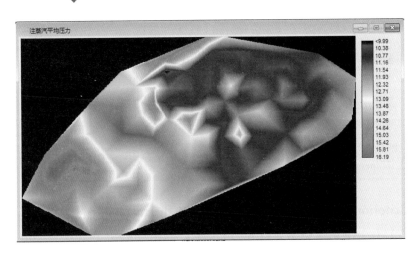

Fig.6-47 Interpolating surface of average pressure

From the background of this reservoir, we can know that it is a steam injection oilfield. Therefore, the maximum risk comes from max injection temperature, then average pressure, volume of steam injection, volume of production and steam injection times. According to analytical hierarchy program (AHP) method, the relative importance between factors can be expressed by a significant value. The comparison matrix is constructed as Table 6-4.

Table 6-4 Comparison Matrix

附录 B	附录 C Injection times	附录 D Volume of production	附录 E Volume of injection	附录 F Average pressure	附录 G Max temperature
附录 H Injection times	附录 I	附录 J /2	附录 K /3	附录 L /5	附录 M /7
附录 N Volume of production	附录 O	附录 P	附录 Q /2	附录 R /3	附录 S /5
附录 T Volume of injection	附录 U	附录 V	附录 W	附录 X /2	附录 Y /3

Continued

附录 B	附录 C Injection times	附录 D Volume of production	附录 E Volume of injection	附录 F Average pressure	附录 G Max temperature
附录 Z Average pressure	附录 AA	附录 BB	附录 CC	附录 DD	附录 EE /2
附录 FF Max temperature	附录 GG	附录 HH	附录 II	附录 JJ	附录 KK

Criterion weights are calculated, and the results are 0.0524, 0.0887, 0.152 4, 0.261 9, and 0.444 6. The risk assessment result is shown as Figure 6-48. From Figure 6-48, it is found that the risk of casing failure increases gradually from southwest to northeast. The dangerous areas are located in the northeast, with risk values large than 200, especially in some part of the oilfield, values even more than 250. Therefore, more attention should be pain on those areas, and more comprehensive measures are needed to reduce the risk. In the western, values are properly less than 100, it reveal relatively safety areas.

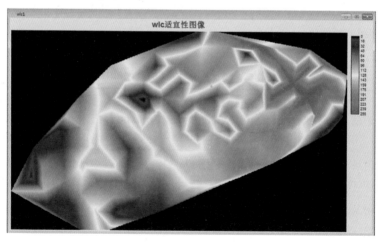

Fig.6-48   Evaluation results

Chapter 6 Application Examples of Geostatistics

Most engineering data exhibit some spatial variability that can be described relative to distance and direction. Spatial variability present in the sample data can be assessed in terms of distance and direction. During spatial variability analysis, it is necessary to spend a great deal of time modeling different directions with many different distances and lags, and confirming results with knowledge about the data distribution.

6.3 Application in Reservoir Heterogeneity

As an example application in reservoir heterogeneity, reservoir properties of Leng43 block in Liaohe Oilfield will be investigated. There are 51 oil wells in this reservoir. The data of oil wells include porosity, permeability, water saturation, effective thickness of strata, and shale percentages, they are shown in Table 6-5. Reservoir heterogeneity will be analyzed based on those data in this exercise.

Table 6-5 Data of reservoir properties

Wells	X-axis	Y-axis	porosity	K	Sw	Thickness	Clay(%)
LENG128	1.2	3.9	11	2.8	40	27	28
19-172	0.2	4.6	10	2.1	42	27	35
22-560	2	1	14	3.2	75	29	20
25-174	2.8	5.2	12.5	4	33	23	51
25-166	3.1	3	12.8	3.6	35	23.5	47
27-176	3.3	6	13.2	3.8	30	26	70
27-168	3.5	3.7	13.1	3.9	40	22	49
29-166	4.5	3	14	4.1	30	24	70
31-172	4.7	5	12.7	4.2	46	25	72
31-168	5.5	3.7	14	4.3	46	30	78

Continued

Wells	X-axis	Y-axis	porosity	K	Sw	Thickness	Clay(%)
32-562	6	2	12.5	4.1	50	35	60
32-554	6.9	0	7.9	2.6	65	44	28
33-174	7.5	5.7	15	3.6	55	37	100
33-170	6.1	4.3	13.2	4.1	45	41	59
36-566	7.5	3.2	13.1	3.6	46	46	68
36-558	8	0.9	11.8	2.8	62	40	39
39-172	8.4	5.2	13.5	3.6	56	37	75
44-570	10.9	4.7	12.6	5.9	62	48	58
48-570	12.4	4.9	13	3.8	55	35	70
48-566	13.1	3.7	12.8	2	52	30	68
GUAN12	13	2.5	12.5	1.9	51	31	52
48-554	13.5	0.3	11.7	2.2	52	28	41
56-664	16.1	3.4	11.5	3.6	57	34	49
56-558	16.5	1.6	11.2	2.6	61	44	30
60-570	17.5	5.4	14	3	53	39	100
60-664	17.8	3.5	12.1	2.8	56	33	49
64-562	19.3	3.1	12.2	3.6	57	30	41
64-558	20	2	12	2.8	48	28	35
64-554	20.3	0.8	9	6	52	31	22
LENG86	21.3	3.5	12.2	8	61	22	37
72-570	22.8	5.9	13.5	3.8	48	32	62
72-566	22.5	4.8	12.8	3.6	49	28	50
72-558	23.1	2.2	11.6	3	48	21	28
76-664	24.7	4.1	12.5	4.2	55	26	21

Contrnued

Wells	X-axis	Y-axis	porosity	K	Sw	Thickness	Clay(%)
76–558	25.2	2.4	11.8	3.5	54	23	38
80–570	25	6.2	12.9	2	50	18	66
80–566	26.3	5.1	14	1.8	40	16	81
80–558	26.7	2.6	12.5	3.2	62	23	49
84–562	28.1	3.9	14.8	3.8	45	14	91
84–554	28.5	1.6	11.9	2.8	65	18	47
88–566	29.2	5.1	16.1	2.9	42	20	100
88–562	30.1	4.1	14.1	4.1	50	15	79
88–558	30.2	3	13.2	5.2	60	17	61
92–566	31.3	5	14.6	3	45	17	100
92–554	32.2	2	12.5	5.6	57	15	50
96–566	32.8	5	14	5.2	46	17	79
96–562	33.4	4.1	12	11	55	14	45
96–558	33.3	3.1	11.8	8	65	13	50
100–558	34.7	3.2	11	7	64	10	36
102–604	35.4	5.1	13.2	4	47	8	50
102–554	35.2	2.2	10.8	10	60	5	31

With the same method as above, spatial variability is analyzed. Because we don not need to evaluate the total risk of reservoir, only interpolating surfaces are needed to be creating to analysis the reservoir heterogeneity. Now we begin this procedure as follow.

6.3.1 The spatial variability of water saturation

The data of water saturation in 51 wells are entered as the input vector variable file, see Figure 6-49. And the semi-variogram surface map is shown as Figure 6-50.

UNI_ID	ID	X	Y	HSBHD
1	LENG128	1.2	3.9	40
2	19-172	0.2	4.6	42
3	22-560	2	1	75
4	25-174	2.8	5.2	33
5	25-166	3.1	3	35
6	27-176	3.3	6	30
7	27-168	3.5	3.7	40
8	29-166	4.5	3	30
9	31-172	4.7	5	46
10	31-168	5.5	3.7	46
11	32-562	6	2	50
12	32-554	6.9	0	65
13	33-174	7.5	5.7	55
14	33-170	6.1	4.3	45
15	36-566	7.5	3.2	46
16	36-558	8	0.9	62
17	39-172	8.4	5.2	56
18	44-570	10.9	4.7	62

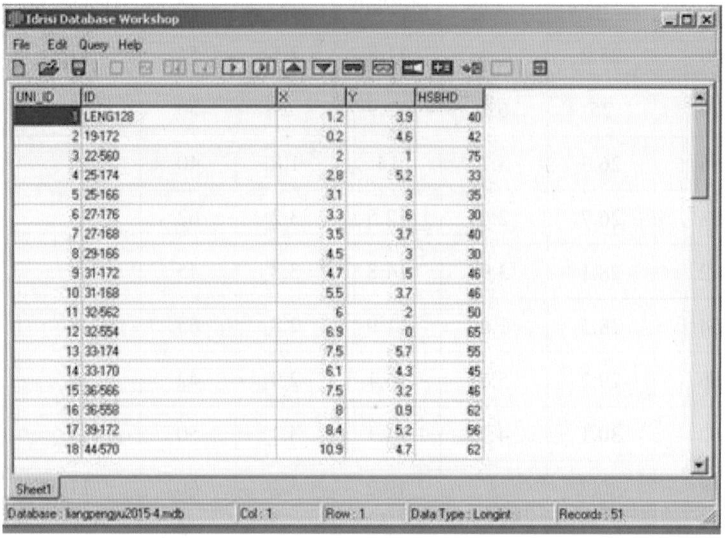

Fig.6-49　Data of water saturation

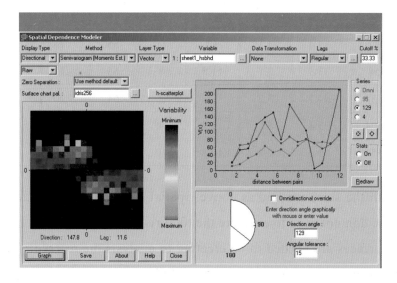

Fig.6-50　Semi-variogram surface of water saturation

The semi-variogram surface image of water saturation is calculated as Figure 6-50. In semi-variogram surface images, lag distance zero located

Chapter 6 Application Examples of Geostatistics

at the center of the grid, lag distances increase outwardly from center in all directions. Thus each pixel represents an approximate average of the pairs' semivariance. Variability increases from low to high with colors change from dark blue to green colors. The direction and the number of lags are displayed in the bottom of the semi-variogram surface image. In the spatial variability of water saturation, the max direction of spatial variability is 129 degree, and the min direction is 95 degree. The omnidirectional series is similar to an average in all directions and therefore it falls between the two series of 129 and 95 degree.

6.3.2 The spatial variability of porosity

With the same procedure, the data of porosity in 51 wells are entered as the input vector variable file, see Figure 6–51. The semi-variogram surface image of porosity is also calculated as Figure 6–51.

UNI_ID	ID	X	Y	KXD
1	LENG128	1.2	3.9	11
2	19-172	0.2	4.6	10
3	22-560	2	1	14
4	25-174	2.8	5.2	12.5
5	25-166	3.1	3	12.8
6	27-176	3.3	6	13.2
7	27-168	3.5	3.7	13.1
8	29-166	4.5	3	14
9	31-172	4.7	5	12.7
10	31-168	5.5	3.7	14
11	32-562	6	2	12.5
12	32-554	6.9	0	7.9
13	33-174	7.5	5.7	15
14	33-170	6.1	4.3	13.2
15	36-566	7.5	3.2	13.1
16	36-558	8	0.9	11.8
17	39-172	8.4	5.2	13.5
18	44-570	10.9	4.7	12.6

Fig.6-50 Data of porosity

From Figure 6-51, it can be found that the max direction of spatial variability of porosity is 63 degree, and the min direction is 100 degree. The trend of the spatial variability of porosity is similar with the spatial variability of water saturation. The value of the spatial variability of porosity is higher than the spatial variability of water saturation.

We can deduce that the similar trend imply the same controlling factor. Thus, the spatial variabilities of porosity and water saturation are controlled by the same factors, and we can guess here it is geological condition. The main difference between Figure 6-49 and Figure 6-51 is the spatial variability near the border.

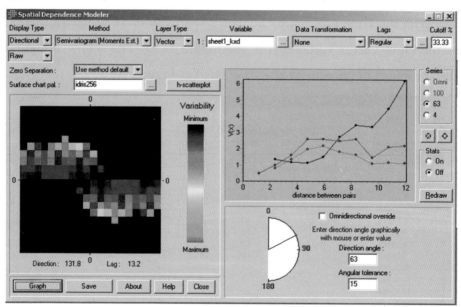

Fig.6-51 Semi-variogram surface of porosity

6.3.3 The spatial variability of permeability

The data of permeability in 51 wells are entered as the input vector variable file, see Figure 6-52.

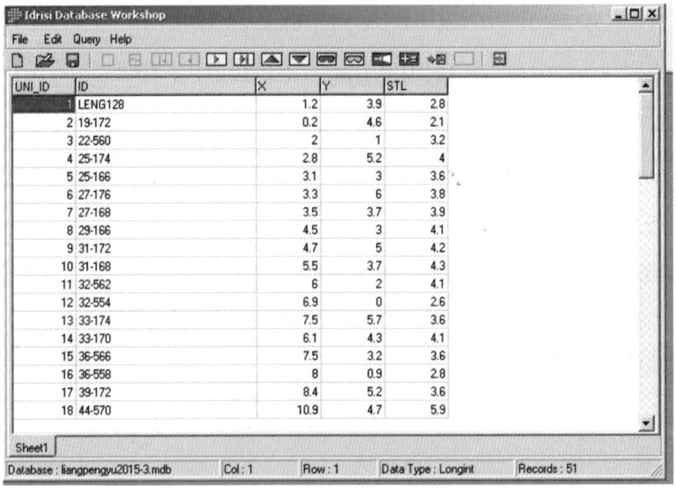

Chapter 6 Application Examples of Geostatistics

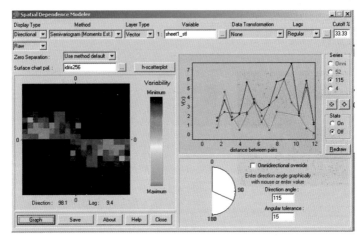

Fig.6-52 Data of permeability

The semi-variogram surface image of permeability is also calculated as Figure 6-53. From Figure 6-53, it can be found that the max direction of spatial variability of porosity is 115 degree, and the min direction is 52 degree. The trend of the spatial variability of permeability is similar with the spatial variability of water saturation and porosity, but with low variability.

Fig.6-53 Semi-variogram surface of permeability

6.3.4 The spatial variability of effective thickness

With the same procedure, the data of effective thickness of sandstone in 51 wells are entered as the input vector variable file, see Figure 6–54. The semi–variogram surface image of porosity is calculated and shown as Figure 6–55.

UNI_ID	ID	X	Y	HD
1	LENG128	1.2	3.9	27
2	19-172	0.2	4.6	27
3	22-560	2	1	29
4	25-174	2.8	5.2	23
5	25-166	3.1	3	23.5
6	27-176	3.3	6	26
7	27-168	3.5	3.7	22
8	29-166	4.5	3	24
9	31-172	4.7	5	25
10	31-168	5.5	3.7	30
11	32-562	6	2	35
12	32-554	6.9	0	44
13	33-174	7.5	5.7	37
14	33-170	6.1	4.3	41
15	36-566	7.5	3.2	46
16	36-558	8	0.9	40
17	39-172	8.4	5.2	37
18	44-570	10.9	4.7	48

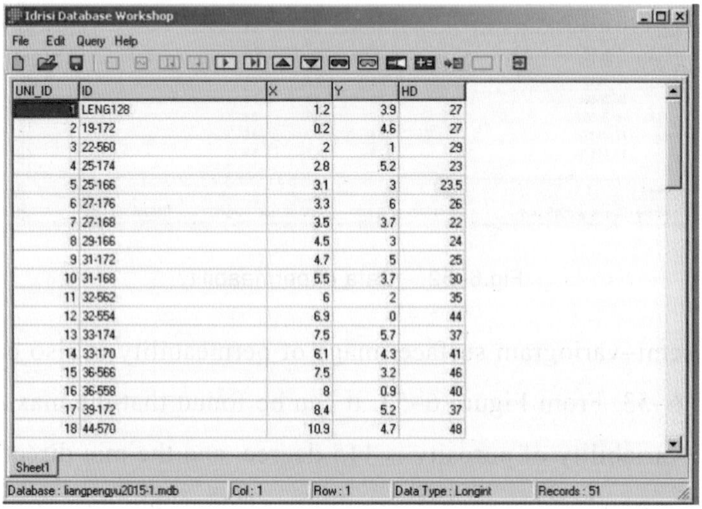

Fig.6–54 Data of effective thickness

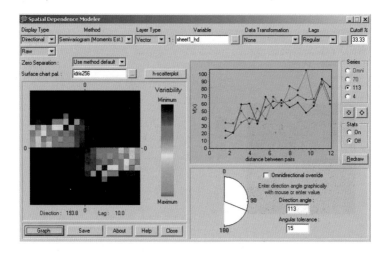

Fig.6–55 Semi–variogram surface of effective thickness

Chapter 6 Application Examples of Geostatistics

For effective thickness of sandstone, the minimum spatial variability direction is distributed between 100 degree and 130 degree, the maximum spatial variability direction is between 40 degree and 90 degree. In direction 100 degree to 130 degree, variability increases rapidly. The omnidirectional series is similar to an average in all directions and therefore it falls between the max and min series.

Above all, it can be found that semi-variogram surfaces of water saturation, porosity, and permeability are similar, this imply that they are controlled by the similar geological condition. Also, the semi-variogram surface of effective thickness of sandstone is different with others. This implies that the controlled geological condition of effective thickness is different with others. These reveal that different geological condition affect the distribution of reservoir property and their spatial variability.

6.3.5 Distribution of reservoir property

In this step, interpolated surfaces of above four factors of reservoir property are created, see Figure 6-56 and Figure 6-59.

Fig.6-56　Distribution of effective thickness

Fig.6-57　Distribution of water saturation

Fig.6-58　Distribution of porosity

Fig.6-59　Distribution of permeability

Four interpolated surface maps for all the four influence factors are worked out. In those maps, the effective thickness creases from east to west, most parts of the research area is low, and only a small high permeability area exist in the east part. The porosity and water saturation changes from south to north, the porosity increases from south to north, but water saturation decreases. Of course, you can make further professional analysis about relations among those factors.

References

[1] Jacek. Malczewski. Ordered weighted averaging with fuzzy quantifiers: GIS-based multicriteria evaluation for land-use suitability analysis [J]. International Journal of Applied Earth Observation and Geoinformation, 2006, 8(4): 270–277.

[2] Yager R R. Quantifier guided aggregation using OWA operation [J]. International Journal of Intelligent Systems, 1996,11(1): 49–73.

[3] Jacek Malczewski. GIS-based land-use suitability analysis: a critical overview[J]. Progress in Planning, 2004,6(2): 3–65.

[4] Liu Guili. The first research on constructive land suitability evaluation in connective location between town and country [J]. Geographical Research, 2000,19(1):80–85.

[5] Chen Songlin. Evaluation on the Suitability of the Uncultivated Land Based on GIS [J]. Fujian Geography, 2001,16(1): 34–37.

[6] Chen Fen. Evaluation on the Land Suitability in Fujian Based on AEZPGIS [J]. Fujian Grography, 2002,17(3): 11–18.

[7] Qiu Bingwen, Chi Tianhe, Wang Qinmin. Et al. Application of GIS and its Prospect in Land Suitability Assessment [J]. Geography and Geo-Information Science, 2004, 20(5):20–23,44.

[8] Li Zhengguo, Wang Yanglin, Wu Jiansheng.et al. Application of Urban Growth Model in Spatial Decision-Making of Land Supply: The Case of Longgang District of Shenzhen City[J]. Resources science, 2005, 27(2):51–58.

[9] Ma Gang, Li Haiyu, Xu Yilun. Analysis on Nanjing urban land use potential based on GIS [J]. Geography and Geo-Information Science, 2005,21(3):56-59.

[10] Qian Huaisui, Ren Yuyu, Li Mingxia. Changes of Cotton Climate Risk Degree in Henan Province [J]. ACTA Geographical SINICA, 2006,61(3):319-326.

[11] Chen Wen, Sun Wei, Duan Xuejun, et al. Regionalization of Regional potential development in Suzhou City [J]. ACTA Geographical SINICA, 2006,61(8): 839-846.

[12] Ou Xiong, Feng Changchun, Shen Qingyun, et al. Application of Synergisticity Model in Urban Land-Use Potential Appraisal [J], Geography and Geo-Information Science, 2007, 23(1): 42- 45.

[13] Xu Yejun, Da Qingli. Trapezoidal fuzzy ordered weighted averaging operator and its application to decision making [J]. Journal of Southeast University (Natural Science Edition), 2006,36(6): 1034-1038.

[14] Liu Yi, Gao Xiaoguang, Lu Guangshan, et al. Weighted Attribute Information Fusion Based on OWA Aggregation Operator [J]. Chinese Journal of Scientific Instrument, 2006,27(3): 322-325.

[15] Zhang Wenhong, Chen Senfa. Research on hybrid index hierarchy fuzzy decision making method [J]. Journal of Management Sciences in China, 2005, 8(1): 7-11.

[16] Ni Xiancun, Zuo Hongfu, Bai Fang, et al. Evaluation of Spare Parts Suppliers Based on Fuzzy Multiple Attribute Group Decision Making and OWA Operators [J]. Mechanical Science and Technology, 2006,25(12): 1404-1407.

[17] Liu Yi, Gao Xiaoguang, Lu Guangshan, et al. Multisensor Target Recognition Based on the OWA Aggregation Operator [J]. Chinese Journal of Sensors and actuators, 2006,19(2): 530-533.

[18] Xu Zeshui, Da Qingli. Hybrid aggregation operator and it s application to multiple attribute decision making problems [J]. Journal of Southeast

University (English Edition), 2003, 19(2): 174–177.
[19] Li Baojun. Improve Integrative Coordination Ability for Disaster Prevention and Reduction in City [J]. Natural Disaster Reduction in China, 2004,(12): 23–24.
[20] Zhu Qingjie, Su Youpo. Research on integrated geoharzard prevention in Tangshan City [J]. Journal of Disaster Prevention and Mitigation Engineering, 2005, 26(3): 309–314.
[21] Zhu Qingjie, Ma Yajie, Chen Yanhua. Evaluation of regional crust based on ANN [J], Chinese Jounal of Geotechnical Engineering, 2005,27(9): 1105–1109.
[22] Zhang Xiuyan, Zhu Qingjie, Wang Zhitao. A Study of Land Exploitative Intensity [J]. Journal of Hebei Institute of Technology, 2007,29(2): 140–143.
[23] Lan Rongxin, etc. The Developing Status and Trends of GIS [J]. Geospacial Information Journal, 2004, 2 (1):9–12.
[24] Meng Jijun. Land Evaluation and Administration[M]. Beijing: Science Book Concern, 2005.
[25] Chen Jing, Deng li, Zhu Qingjie. Management information system for integrated disasters prevention in Tangshan city[J]. World Earthquake Engineering, 2005, 21(1): 66–69.
[26] Zhou Lijun, Zhang Shuhua, Zang Shuying. The Application of Spatial Analyst in Sites Selection — A Case of Daqing City in the Ha–Da–Qi Industrial Corridor [J]. Areal Research and Development, 2007, 26(1): 125–128.
[27] Liu J.2003. Application of AHP in comprehensive assessment of water environment quality [J]. Journal of Chongqing Jianzhu University, 25(1): 77–81.
[28] Liu XW. 2004. Three methods for generating monotonic OWA operator weights with given orness level [J]. Journal of Southeast University (English Edition), 20(3): 369–373.

[29] Makropoulos CK, Butler D. Spatial ordered weighted averaging: incorporating spatially variable attitude towards risk in spatial multi-criteria decision-making [J]. Environmental Modelling & Software, 2006.,21 (1): 69-84.

[30] Wang YM, Luo Y, Liu XW.Two new models for determining OWA operator weights[J]. Computers & Industrial Engineering, 2007, 52 (2): 203-209.

[31] Y.M. Wang, Y.Luo, and Z.S. Hua, "Aggregating preference rankings using OWA operator weights," Information Sciences, Vol. 177, No. 16, 2007, pp.3356-3363.

[32] R.R.Yager, "Using trapezoids for representing granular objects: applications to learning and OWA aggregation," Information Sciences, Vol. 178, No. 1, 2008,pp.363 – 380.

[33] D.Wu, "Performance evaluation: an integrated method using data envelopment analysis andfuzzy preference relations," European Journal of Operational Research, Vol. 194, No. 1, 2009, pp.227-235.

[34] Q.J. Zhu, Y.P. Su, and D.D. Wu, "Risk assessment of land-use suitability and application to Tangshan City," Int. J. Environment and Pollution, Vol. 42, No. 4, 2010, pp.330-343.

[35] Yager R.R. On ordered weighted averaging aggregation operators in multi-criteria decision making [J]. IEEE Transactions on Systems, Man Cybernetics, 18 (1) : 183-190, 1988.

[36] J.Figueira, Greco, M.Ehrgott. Multiple Criteria Decision Analysis：State of the Art Surveys[M]，Springer-Verlag，NewYork，2005.

[37] V. Belton, T.J.Stewart. Multiple Criteria Decision Analysis：An integrated approach[M], Boston, Klumwer Academic Publishers, 2002.

[38] Simon H.A，The New Science of Management Decision[M]，Harper and Brothers，1960.

[39] Byeong S A. Preference relation approach for obtaining OWA operators weights. International Journal of Approximate Reasoning, 2008, 47

(2):166-178.

[40] Rahman, M.T, Rashed. Towards a Geospatial Approach to Post-Disaster Environmental Impact Assessment. Proceedings of the 4th International Conference on Information Systems for Crisis Response and Management ISCRAM 2007, 2007, 219-226.

[41] Claus Rinner. Multi-Criteria Evaluation in Support of Emergency Response Decision- Making, Proceedings of Joint CIG/ISPRS Conference on Geomatics for Disaster and Risk Management, Toronto, 2007: 23-25.

[42] Xu Z S, Chen J. An interactive method for fuzzy multiple attribute group decision making. Information Sciences, 2007, 177 (1): 248-263.

[43] B.S. Ahn, On the properties of OWA operator weights functions with constant level of orness, IEEE Transactions on Fuzzy Systems 14 (2006) 511-515.

[44] Xinwang Liu. On the maximum entropy parameterized interval approximation of fuzzy numbers[J], FUZZY SETS AND SYSTEMS, 2006, 157: 869-878.

[45] Michael F.Goodchild. GIS and disasters: Planning for catastrophe[J], Computers, Environment and Urban Systems, 2006 (30):227-229.

[46] C. Rinner, M. Raubal, Personalized multi-criteria decision strategies in location-based decision support[J], Journal of GeographicInformation Sciences, 2004, 10: 149-156.

[47] Ha-Rok Bac, Ramana V,Grandhi etal. Epistemic Uncertainty Quantification Techniques including evidence theory for large scale structures[J], computers structures, 2004(82):1101-1112.

[48] V. Torra, OWA operators in data modeling and reidentification, IEEE Transactions on Fuzzy Systems 12 (2004), 652-660.

[49] Yager R R, Induced aggregation operators[J], Fuzzy Sets and Systems, 2003(137): 59-69.

[50] Z.S. Xu, Q.L. Da, An overview of operators for aggregating information,

International Journal of Intelligent Systems[J], 2003, 18, 953-969.

[51] Tarek Rashed, John Weeks, Assessing vulnerability to earthquake hazards through spatial multicriteria analysis of urban areas[J], International Journal of Geographical Information Science, 2003, 17 (6): 547-576

[52] S.Chakhar, J.-M.Martel. Enhancing geographical information systems capabilities with multi-criteria evaluation functions Journal of Geographic Information and Decision Analysis, 2003, 7(2): 47-71

[53] Riberito R A, Pereita R A M. Generalized mixture operators using weighting functions:a comparative study with WA and OWA[J], European Journal of Operational Research, 2003, 145 :329-342

[54] C. Rinner, J. Malczewki, Web-enabled spatial decision analysis using ordered weighted averaging (OWA) [J], Journal of Geographical Systems, 2002, 4: 385-403

[55] R. Fuller, P. Majlender, An analytic approach for obtaining maximal entropy OWA operator weights[J], Fuzzy Sets and Systems 124, (2001) 53-57

[56] Herrera F,Herrera-ViedmaE,ChiclanaF. Multiperson decision-making based on multiplicative preference relations[J], European Journal of Operational Research ,2001, 129: 372-385.

[57] Despic O., Simonovic S., Aggregation operators for soft decision making in water resources[J], Fuzzy Sets and Systems, 2000, 115(1), 11-33

[58] H. Jiang, J.R. Eastman, Application of fuzzy measures in multi-criteria evaluation in GIS[J], International Journal of Geographical Information Science, 2000, 14: 173-184

[59] Eastman J R, Jiang H and Toledamo J, Multi-criteria and multi-objective decision making for land allocation using GIS, In "Multicriteria anlysis for land-use management" ed. by Beinat E, and

References

Nijkamo P. (Kluwer Academic Publishers), 1998,227-251.

[60] D. Filev, R.R. Yager, On the issue of obtaining OWA operator weights, Fuzzy Sets and Systems, 1998, 94, 157-169

[61] Goodchild M F, Hunter G J, A simple positional accuracy measure for linear features [J], International Journal of Geographical Information Science, 1997, 11: 299-306

[62] F. Herrera, E. Herrera-Viedma, J.L. Verdegay, Direct approach processes in group decision making under linguistic OWA operators[J],Fuzzy Sets and Systems, 1996, 79, 175-190

[63] Mario Mejia Navarro, Luis A. Garcia. Natural hazard and risk assessment using decision support systems, application; Glenwood Springs[J], Colorado Environmental and Engineering Geoscience, 1996, Vol.2（3）: 299-324

[64] J. Malczewski, Integrating geographical information systems and multiple criteria decision-making methods[J], International Journal of Geographical Information systems, 1995, 9(3):251-273.

[65] D. Filev, R.R. Yager, Analytic properties of maximum Entropy OWA operators[J], Information Sciences 85 (1995) 11-27

[66] R.J.Eastman, P.A.K.,Kyen, J.Toledno. A Procedure for Multiple-objective Decision Making in GIS under Conditions of Conflicting Objectives[J], Fourth European Conference on GIS ESIG, 93 Proceedings, 1993, vol1, 438-447

[67] Yager R.R, Families of OWA operators[J], Fuzzy Sets and Systems, 1993(59):125-148

[68] Carver SJ, Integrating multi-criteria evaluation with geographical information system[J], International Journal of Geographical Information Systems, 1991, 5(3):321-339

[69] Janssen R, Rietveld P. multi-criteria analysis and GIS:an appliction to agricultural landuse in the Netherlands[J], Geographical Information Systems for Urban and Regional planning, 1990:129-139

[70] M. O'Hagan, Using maximum entropy-ordered weighted averaging to construct a fuzzy neuron, in: Proceedings of 24th Ann. IEEE Asilomar Conference Signals System Comput, Pacific Grove, CA, 1990, 618-623

[71] Roy B. Main sources of inaccurate determination, uncertainty and imprecision in decision models[J], Mathematical and Computer Modeling,1989, 10-11:1245-1254.